PLANT TISSUE CULTURE
(THEORY AND PRACTICALS)

2nd Edition

Other related books

Title	Author
21st Century Biology and Agriculture: Textbook	*A. Raman*
A Text Book of Environmental Science	*Vidya Thakur*
Agricultural Microbiology	*R.P. Pareek*
An Introduction to Fungi 4th Ed	*H.C. Dube*
Analytical Methods for Medicinal Plants and Economic Botany	*M. Daniel*
Applied Microbiology (Agriculture, Environmental, Food and Industrial Microbiology)	*S.M. Reddy*
Botany for NEET and Other Entrance Examinations	*Khwaja Salahuddin*
Colour Atlas of Medicinal Plants	*R.K. Bhutiya*
Essentials of Virology	*S.Ram. Reddy*
Gene Editing : Principles and Applications	*B.D.Singh*
Growth and Development in Plants (21st Century Biology and Agriculture: Textbook Series)	*K.V. Krishnamurthy*
Indiras Objective Agricultural Biotechnology 2nd Ed	*R.L. Arya*
Introduction to Agricultural Botany	*M Pandiyan*
Laboratory Manual of Microbiology	*Vivek Kumar*
Laboratory Manual of Microbiology, Biochemistry and Molecular Biology	*Jyoti Saxena*
MCQ in Agricultural Microbiology	*D. G.Panpatte*
MCQ'S in Microbiology: Advanced	*Balaram Mohapatra*
Medical Microbiology and Immunology	*S.M. Reddy*
Molecular Plant Breeding	*B. D. Singh*
Mycology and Microbiology: A Textbook for UG and PG Courses	*C. Manoharachary*
Objective Botany: MCQ	*D.A. Patil*
Physiology of Crop Plants	*F.P. Gardner*
Physiology of Fungi 3rd Ed	*K.S. Bilgrami*
Plant Biotechnology (A Text Book)	*Sm Paul Khurana*
Plant Biotechnology and Molecular Biology: A Laboratory Manual	*M.S. Punia*
Plant Breeding in 21st Century	*B. D. Singh*
Plant Breeding Theory & Practice (3rd En. Ed)	*N.C. Stoskopf*
Research Methodology in Plant Science 2nd Ed	*P.S. Narayana*

PLANT TISSUE CULTURE
(THEORY AND PRACTICALS)

2nd Edition

T. Pullaiah
Department of Botany,
Sri Krishnadevaraya University,
Anantapur - 515003, Andhra Pradesh (India)

M.V. Subba Rao
Department of Botany,
Andhra University,
Visakhapatnam - 530003, Andhra Pradesh (India)

and

E. Sreedevi
Department of Botany,
Government Degree College,
Anantapur - 515001, Andhra Pradesh (India)

SCIENTIFIC®
PUBLISHERS

Published by:

SCIENTIFIC PUBLISHERS (INDIA)
5A, New Pali Road, P.O. Box 91
Jodhpur 342 001, India
E-mail: info@scientificpub.com
Website: www.scientificpubonline.com

Print, 2026

ISBN: 978-93-86347-34-3
eISBN: 978-93-87869-80-6

Printed in India

PREFACE

Since the publication of the first edition in 2009, several new and exciting developments have taken place in the field of plant tissue culture which pursues us to attempt to provide current information and examples. From this book, faculty in various fields of plant sciences, graduate students, and undergraduates may gain basic and advanced knowledge about application of various aspects of Plant Cell and Tissue Culture. The revised edition presents updated information and will prove to be an excellent text and reference book for theoretical, practical and applied aspects of plant tissue culture with hands on training approach. Each chapter as before, is written in lucid language which has now been thoroughly revised for recent developments to include relevant media protocols, and is profusely illustrated with self-explanatory diagrams. This book includes new additions in every chapter starting from tissue culture history to haploid production, suspension cultures, nucellus culture, protoplast fusion and hybridization and secondary metabolite production and enables to make the readers to know about recent developments in Plant tissue culture. This revised edition includes detailed drawings that complement both knowledge of theory and experiments.

T. Pullaiah
M.V. Subba Rao
E. Sreedevi

LIST OF ABBREVIATIONS

%	Per cent
2iP	2-Isopentenyl adenine
ABA	Abscisic Acid
AC	Activated Charcoal
AIDS	Acquired Immuno Deficiency Syndrome
AR	Analytical Reagent
ATP	Adenosine Triphosphate
B_5	Gamborg's Medium
BA	6-Benzyladenine
BAP	6-Benzyl-Aminopurine
BM	Basal Medium
BuDR	BromoDeoxyuridine
CCC	Cycocel
cDNA	complementary DNA
CH	Casein Hydrolysate
CM	Coconut Milk
CMS	Cytoplasmic Male Sterility
CPPU	N-2-chloro-4puridyl-N-phenyl Urea
CPW	Cell-protoplast Washing
DES	Diethyl Sulphate
DFR	Dihydroflavonol-4-reductase
DH	Diploid Homozygous
DMSO	Dimethyl Sulphoxide
DNA	Deoxyribonucleic Acid

DPU	N,N'-diphenyl Urea
DRT	Decimal Reduction Time
ECs	Embryogenic Clumps
EDTA	EthylenediamineTetraacetic acid
ELISA	Enzyme-linked Immunosorbent Assay
EMS	Ethyl Methane Sulphonate
F3'5'H	Flavanoid 3', 5'- hydroxylase
FDA	Fluorescein Diacetate
FudR	5-Fluorodeoxyuridine
GA_3	Gibberellic Acid
GCMS	Gas Chromatography Mass Spectrometer
GR	Guaranteed Reagent
H_2O_2	Hydrogen peroxide
HEPA	High Efficiency Particulate Air filter
$HgCl_2$	Mercuric chloride
HPLC	High Performance Liquid Chromatography
HU	Hydroxy Urea
IAA	Indole-3-Acetic Acid
IBA	Indole-3-Butyric Acid
IBPGR	International Board for Plant Genetic Resources
IEDCs	Induced Embryogenic Determined Cells
JA	Jasmonic Acid
JAMe	methyl Jasmonate
KN	Kinetin
LN	Liquid Nitrogen
LOS	Low-oxygen Storage
LPS	Low-pressure Storage
LS medium	Linsmaier and Skoog medium
MCO	Mitotic Crossing Over
MCPA	2-methyl-4-chlorophenoxyacetic acid
mRNA	messenger RNA

MS	Murashige and Skoog's Medium
MSO	Methionine Sulphoximide
NAA	Naphthalene Acetic Acid
NOA	Naphthoxyacetic Acid
NR	Nitrate Reductase
^{o}C	Degree Celsius
OD	Optical Density
ODAP	β-oxalyl-L-α,β-diaminopropionic acid
PCPA	4-Chlorophenoxyacetic acid
PCR	Polymerase Chain Reaction
PCV	Packed Cell Volume
PE	Plating Efficiency
PEDCs	Pre-embryogenic Determined Cells
PEG	Polyethylene Glycol
PGRs	Plant Growth Regulators
PLRV	Potato Leaf-Roll Virus
ppm	Parts per million
Psi	Pounds per square inch
PVP	Polyvinylpyrrolidone
RBC	Red Blood Cell
RFLP	Restriction Fragment Length Polymorphism
RO	Reverse Osmosis
rRNA	Ribosomal Ribonucleic Acid
T-DNA	Transfer Deoxyribonucleic Acid
TDP	Thermal Death Point
TDT	Thermal Death Time
TDZ	Thiadiazuron
TMV	Tobacco Mosaic Virus
UV	Ultraviolet
WBC	White Blood Cell
WM	White's Medium

WPM	Woody Plants Medium
TTC	2,3,5-Triphenyltetrazolium Chloride
2,4,5-T	2,4,5-Trichloro phenoxy acetic acid
Picloram	4-Amino, 3,5,6-Trichloropicolinic acid
B-9	Succinic acid 2, 2-dimethyl hydrazide
2,4-D	2,4-Dichloro phenoxy acetic acid

CONTENTS

1 HISTORY OF PLANT TISSUE CULTURE

1.1 INTRODUCTION

The whole gamut of new discoveries in biology and allied sciences can be grouped together under a single term of 'Biotechnology'. It can be defined as a manipulation of biological systems via modern technology to solve practical problems in several fields. It includes clonal propagation via tissue culture, protoplast fusion, gene cloning, DNA recombination, genetic engineering, mutant induction, *in vitro* genetic selection, molecular marker-aided genome analysis and gene mapping (Huang et *al.,* 1993). Enhanced production of pharmaceutically-important secondary metabolites *in vitro* can also be included here.

A large number of techniques developed over the last 150 y have actually laid the foundation of the modern science in agriculture. Development in agriculture since the last 50 y has been greatly aided, in general, by the development of biotechnology, and 'Plant Tissue Culture' and 'Genetic Engineering' in particular (Brar and Khush, 1994).

Since plants are immovable, environment has a very strong impact on every aspect of plant form and function. As a consequence, plant development shows a high degree of plasticity. Environment stimuli are used as cues to rapidly modify and influence the developmental programming to provide flexibility throughout the growth cycle of plant. The entire technology of plant tissue culture is based on this ability of the plant cells to be influenced by their surroundings and differentiate to give rise to a range of organs depending on the culture conditions (Bowles and Leyser, 1994). Plant tissue culture includes the maintenance and propagation of the plant parts, as small as a single cell, in axenic culture under controlled environmental conditions (Pauls, 1995). Plant tissue culture is the technique of growing plant cells, tissues and organs in an artificially-prepared nutrient medium, static or liquid, under aseptic conditions. It is also called *in vitro* culture. The term 'tissue culture' is actually a misnomer; borrowed from the field of animal tissue culture. It is a misnomer because plant micropropagation is concerned with the whole plantlet and not just isolated tissues, though the explant may be a particular tissue. The terms 'plantlet culture' or 'micropropagation', therefore, are more accurate. However, the process remains the same whether we call it cloning, tissue culture, micropropagation, or growing *in vitro*.

The main advantage of tissue culture technique lies in the production of high-quality and uniform planting material that can be multiplied on a year-round basis under disease-free conditions anywhere irrespective of season and weather. Plant cell and tissue culture is fundamental to most aspects of plant biotechnology and is great interest to molecular biologist, plant breeder and industrialists. The tissue or cells obtained from parts of the plants such as stem, root, leaf, shoot apex, etc. are induced to proliferate and produce a new plant. Plant tissue culture falls under following two categories:

1. **Organized growth:** It occurs when plant organs such as apical meristems of the shoots or roots, leaf initials, young flower buds etc. are cultured; when they grow with their preserved tissues or when structures are formed during cultures of previously-undifferentiated tissues. The process is called organogenesis and morphogenesis.

2. **Unorganized growth:** In this type, the tissue formed is an undifferentiated mass of cells, which is called callus. It can be maintained as such for longer periods on media and can be increased in volumes by subculturing.

The different types of plant tissue culture techniques that are mainly based on explants include:

- **Callus tissue culture:** These are cell aggregates that arise from unorganized growth of the small plant parts.

- **Suspension cell culture:** In this type, clumps of the plant cells are dispersed in an agitated liquid media.

- **Protoplast culture:** Culture of the plant cell isolated without cell wall.

- **Anther culture:** This includes the culture of complete anthers containing immature microspores.
- **Organ culture:** It includes culture of leaf primordia, immature flower or fruit. The most important are meristem or shoot tip cultures, single node cultures of separate lateral buds, and embryo culture.

1.2 TOTIPOTENCY OF THE CELLS

Totipotency is the ability of a cell to generate a new organism or part. In plants, all the nucleated body cells have the potentiality to give rise to the whole plant. The ability of a cell or mass of cells to give rise to whole plant is called totipotency or cellular totipotency. The term 'totipotency' was coined by Morgan in 1901 to denote the regeneration of an entire organism from a single cell. Tissue culture works on the principle of totipotency that when suitable conditions are given, even a single cell can produce a whole organism. A zygote, a seed, a spore, a callus, a plant cutting, dedifferentiated cells etc. are totipotent cells. They can develop and differentiate into a new organism or a part.

A callus is totipotent and it grows into a plant. When suitable conditions are provided to the body tissues, the cells grow into callus by means of dedifferentiation (formation of undifferentiated cells from differentiated cells). When undifferentiated cells differentiate into different cell types, the callus undergoes morphogenesis to form shoots and roots, which leads to the formation of an entire plant. This is called redifferentiation.

In animals, zygote alone is a totipotent cell, which is formed by the fusion of male gamete and female gamete cells. It can develop into entire organism consisting of different tissues, organs and organ systems. However, any one cell of the younger one cannot give rise to other younger ones.

During development in the animals, the zygote divides and gives rise to embryonic cells. After a few divisions in embryonic cells, a mass of pluripotent cell is formed. Pluripotent cells can give rise to all tissue types, but they cannot give rise to an entire organism. Thus, pluripotent cells have a very limited potency and they can become one of the several types of cells within a given organ. Haemopoietic blood cells are multipotent and can develop into RBC, WBC, and platelets of the circulatory system. They cannot develop into nerve cells or heart cells.

Pluripotent cells undergo further specialization into unipotent cells. These cells are committed to give rise to cells that have a particular function.

The phenomenon of totipotency is demonstrated with the following experiment: Slices of the carrot root are cut and small pieces of the tissues are taken from the phloem region. These pieces are inoculated into a liquid medium in special flasks, which are rotated slowly. The tissue grows actively, and single cells and small cell aggregates dissociate into the medium (a single cell and some cell aggregates are drawn near the flask). Some of the cell clumps develop roots and when transferred to a semi-solid medium, these rooted nodules form shoots. These plants could be transferred to the soil where they develop into flowering plants. The phloem tissues taken from the roots of these plants can be used to repeat the cycle.

1.3 MORPHOGENESIS

Biological organization of any life is coordinated with several events as though a craftsman was moulding it according to a plan. In this process, the individual parts do not develop independently, but all are knit together into an organized system. The biological science concerned with this dynamic and casual aspect of organic form is called 'Morphogenesis'. The derivation of this word is obvious, the origin of form. It attempts to expose the effects of various factors and how these factors manifest an organic form *in toto*. *'Morphogenesis', a distinctive aspect of organization of life, is the crossroad where all the highways of biological exploration tend to converge'*, says Sinnott. More studies have been made to understand morphogenetic problems of the animals rather than the plants. Recent developments in the plant cells, tissues and organs of higher plants in culture are making the science of plant morphogenesis a fruitful one. Working with plants has a number of advantages. In plants, embryonic regions such as meristem and cambium are permanently available for the study of development. The determinate type of development and abundance of the organs such as leaves, flowers and fruits make the study possible under a wide range of environmental conditions. The behaviour of the individual cells during development differs in plants from that of the animals. In animals, the individual cells are free to move whereas this mobility is absent in the plants, and the cells are almost always attached firmly to the neighbours. Hence, the morphogenetic movements do not have any part in the development. This makes the study of morphogenetic problems simple in plants. The lesser plasticity of the plant cells, their stationary habit, susceptibility to changes under environmental influences, ability to maintain polarity, and differentiation and generation potential make the study relatively simple one. In the field of plant morphogenesis, the contributions made by the scientists such as Hanstein on meristem, Winkler on chimeras, Haberlandt on hormones, Kuster on abnormal growth, Klebs on the effects of the environment and Goebel on the organography are noteworthy. Vochting (1878) stated in his 'Organbildung im Pflanzenseich' that phenomenon of morphogenesis depends on the factors such as polarity, differentiation and regeneration of individual cells, and concluded that the fate of a cell is a function of its position.

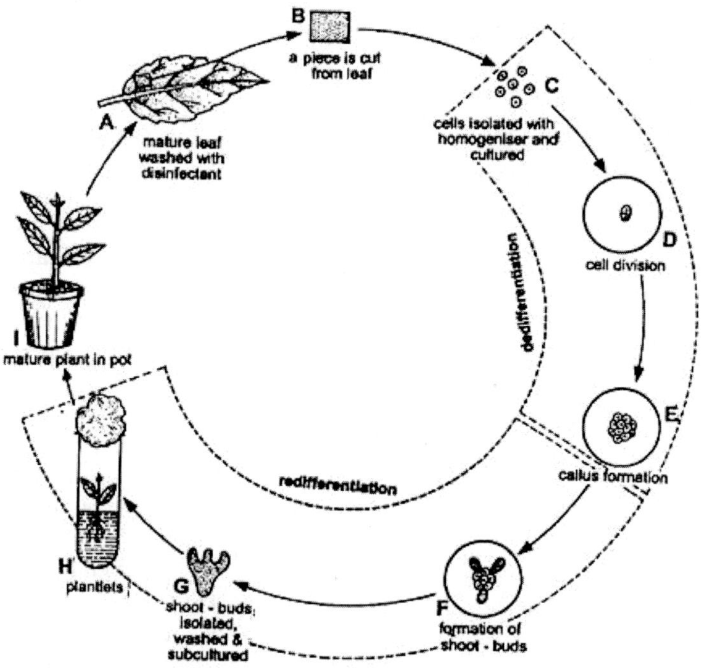

Fig. 1 Scheme depicting the totipotency of the cells

1.3.1 *In Vitro* Morphogenesis

Under normal conditions, a seed – the miniaturized sporophyte – has the message to reconstitute an entire plant with similar shape, structure and function of the mother plant. All known about this phenomenon is that a complex adult multicellular organism has emerged from a relatively-simple organized zygote through a sequence of mitosis (Fig. 1). This *de novo* origin of structures and functions from a fertilized egg or zygote is a complicated phenomenon in which most of the events are not yet known in detail. This is the state in plants *in vivo* or in an entire plant. Considering the cells or organs cultured *in vitro*, the morphogenesis is still an event without many details. However, in the aspects of *in vitro* morphogenesis, significant progress has been made after the discovery of totipotency of plant cells, phytohormones, and the hypothesis of regulation of morphogenesis by the critical balance between auxin and cytokinin.

Various terms are used to define the phenomenon in *in vitro* studies. For example, differentiation, dedifferentiation, redifferentiation, regeneration and morphogenesis are the terms with overlapping meanings. The sharp differences among these terms are exposed hereunder to give a clear-cut view for their usage:

- **Differentiation:** The term 'differentiation' is used in many different senses in biology. In broad sense, it is defined as the process by which meristematic cells are converted into two or more types of cells, tissues or organs, which are qualitatively different from each other.
- **Dedifferentiation:** This term is used to denote the process of formation of unorganized tissues from the highly organized tissues.
- **Redifferentiation:** It is the process of differentiation occurring in an undifferentiated tissue.
- **Regeneration:** It is defined as the structuring of any part, which has been removed or physiologically isolated from the organism. In other words, genesis of an entire plant from the cultured explants directly or indirectly via callus is called regeneration.
- **Morphogenesis:** Attainment of a biological organization or form is termed as morphogenesis. Under *in vitro* conditions, this can be achieved by two routes: 1. *de novo* origin of organs either shoots or roots, from the cultured tissues precisely termed as 'organogenesis' and 2. *de novo* origin of embryos with distinct root and shoot poles on opposite ends from the somatic cells or cells cultured *in vitro*, otherwise called as somatic embryogenesis. The historical background, achievements and the causes for the two routes are discussed below.

- **Organogenesis:** In plant tissue culture, organogenesis means genesis of organs such as shoots, roots, leaves, flowers, etc. The earliest report on induction of shoot organogenesis *in vitro* was made by White (1939) using a tobacco hybrid; and the first observation of root formation was reported by Nobecourt (1939) using carrot callus. The basic regulatory mechanism underlying in organogenesis was not identified till late 1950s. Skoog and Miller (1957) were responsible to recognize the regulatory mechanism as a balance between auxin and cytokinin. As per their finding, a relatively high level of auxin to cytokinin favoured root formation and the reverse favoured shoot formation. Using this concept, it has now become possible to achieve organogenesis in a large number of plant species by culturing explants, calli and cell suspension in a defined medium.

In organogenesis, the shoot or root may form first depending upon the nature of growth hormones in the basal medium. The genesis of shoot and root from the explants or calli is termed as 'caulogenesis' (caulm = stem) and 'rhizogenesis' (rhizo = root) respectively (Fig. 2). It is a general rule that the organ formation would be through a process of differentiation in the undifferentiated mass of parenchyma. Most of the parenchymatous cells are highly vacuolated, and are with inconspicuous nuclei and cytoplasm; sometimes with lignification. In this group of cells, regions showing random cell division would occur, leading to radial files of differentiated tissues. These scattered cell-division regions would form regions of high mitotic activity resulting in the formation of meristematic centres, otherwise termed as 'meristemoids'.

These meristemoids may be either on the surface of the calli or embedded in the tissue. Continued cell division in these meristemoids would produce small protuberances on the surface of the calli, giving a nodular appearance to the tissues. The primordia of organs from the meristemoids form either the shoot or root by repeated mitotic activity. This was discovered by Torrey in 1966. The meristemoids consist of a spherical mass of small isodiametric meristematic cells with dense cytoplasm and a high nucleo-cytoplasmic ratio. Normally, callus tissues accumulate starch and other crystals before organogenesis, but the substances disappear during meristemoid formation. During the initial stages of meristemoid formation, the cytoplasmic protrusions enter the vacuoles, thus distributing the vacuoles around the periphery of each cell or dispersed throughout the cytoplasm. The nucleus is in the centre with maximum possible size. Thus, the cells in the meristemoids resemble the cells of highly-active meristem in an intact plant.

- **Embryogenesis:** An embryo is defined as a plant in its initial stage of development. Each embryo possesses two distinct poles; one to form root and the other to form shoot, and is the product of fusion of gametes. In some plant species, embryos are produced without the fusion of gametes and termed as 'asexual embryogenesis' or 'adventitious embryony'. In an intact plant, this type of embryogenesis may occur in sporophytic tissues such as integuments and nucellar tissues or from unfertilized gametic cells.

Apart from the normal course of embryo formations *viz.*, zygotic embryogenesis and adventitious embryony, instances of embryo formations from the tissue cultures *in vitro* were reported. This phenomenon termed as 'somatic embryogenesis' was first observed by Steward and his coworkers (1958) in suspension cultures of carrot followed by Reinert (1959). Since then, a number of reports of embryo formation have been published.

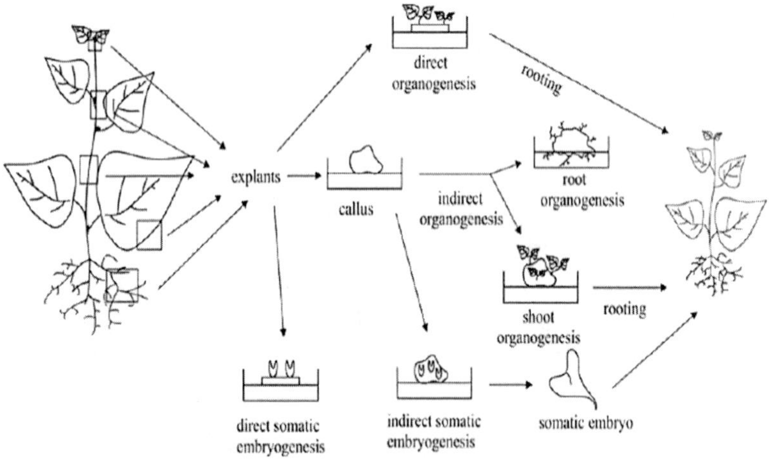

Fig. 2 Organogenesis and plantlet development

Somatic embryogenesis or *in vitro* embryogenesis produces embryo-like structures resembling the zygotic embryos in structure and morphogenetic potential. Despite this resemblance, the ontogeny of an embryo-like structure from somatic cell differs from that of zygotic embryo, where the origin is from a single cell. Embryoid is generally used to denote the embryo-like structure from cultured tissues. These embryoids possess bipolarity, no vascular connection with the mother tissue and origin from a single cell or a group of cells.

1.4 HISTORY OF PLANT TISSUE CULTURE

Henri-Louis Duhamel du Monceau's (1756) pioneering experiments on wound healing in plants demonstrated spontaneous callus formation on the decorticated region of the elm plants. His studies, according to the noted biologist Gautheret (1985), could be considered a 'foreword' for the discovery of plant tissue culture.

The theoretical basis of tissue culture, in general, was established as early as 1838–1839 in Schleiden and Schwann's Cell Theory. This was followed by the work of Trecul (1853) who performed his experiments on callus formation by decorticated trees such as *Robinia, Pawlonia* and *Ulmus*. Vochting (1878) through his experiments on *Brassica rapa*, obtained very luxuriant callus, and also proposed the polarity in the development of buds from the upper portion and roots or callus from the lower portion of a stem piece. Rechinger (1893) described the formation of callus on the isolated stem fragments as well as on the root slices. He placed segments of poplar stem and pieces of beet and candeloin roots on a wet filter paper, and observed callus formation.

It was Thomas Hunt Morgan (1901) who coined the term, 'totipotency' to describe the cell's capability to form an individual plant. Gottlieb Haberlandt, a German botanist, in 1902 first attempted to prove this theory experimentally using monocotyledonous plants. Though his attempt was unsuccessful, he elaborated the concept of totipotency that refers to the potential of an individual cell to regenerate into a whole plant. With the passage of time, most of the predictions in his paper written in 1902 were confirmed experimentally, proving Haberlandt's broad vision and foresight. Hence, Haberlandt is regarded as the 'Father of Plant Tissue Culture'. Haberlandt started his experiments in 1898 using single cells isolated from the palisade tissue of the leaves of *Lamium purpureum,* pith parenchyma from petioles of *Eicchornia crassipes*, epidermis of *Ornithogalum,* stamen hair cells of *Tradescantia* and glandular hair of *Pulmonaria* and *Urtica*. He grew them on Knop's salt solution with sucrose, and observed obvious growth in the palisade cells. However, unfortunately none of the cells divided, but this did not deter others to perform experiments.

In 1904, Hannig had initiated a new line of investigation, which later developed into an important applied area of *in vitro* techniques. Hannig (1904) cultured nearly mature embryos from the seeds of several species of crucifers (*Raphanus sativus*, *R. landra*, *R. caudatus* and *Cochlearia danica*). Laibach (1925, 1929) demonstrated the practical application of embryo culture in the field of plant breeding. Carrel (1910) was the first scientist who demonstrated the culture of living cells outside the body of an organism.

Haberlandt's lack of success was not in vain because one of his students, Kotte (1922) from Germany, reported the growth of isolated root tips on a medium consisting of inorganic salts. At the same time, Robbins reported similar success with root tips of maize. Although these root cultures could not be maintained indefinitely, they proved that it was easier to maintain the cultures having meristematic cells. The first successful report of continuously-growing cultures of tomato root tips was made by White (1934). Initially, White used a medium containing inorganic salts, yeast extract and sucrose, but later, yeast extract was replaced by three B-vitamins *viz.*, pyridoxine, thiamine and nicotinic acid (White, 1937). Later on, Gautheret (1939) established continuously-growing tissue cultures from carrot root cambium. In the same year, White (1939) reported the establishment of similar cultures from tumour tissue of the hybrid *Nicotiana glauca* × *N. langsdorffii*. Finally, the possibility of cultivating the plant tissues for unlimited period using a medium enriched with auxins was announced independently by Gautheret, White and Nobecourt (1939). Experiments on crown gall and tumour formation in the plants were initiated by White and Braun (1942).

Van Overbeek *et al.* (1941), for the first time, reported the stimulatory effect of CM on the embryo development and callus formation of *Datura*. The combination of CM and 2,4-D had a remarkable effect on the proliferation of the cultured carrot and potato tissues (Steward and Caplin, 1951). The experiments on crown gall and tumour formation in the plants were initiated by White and Braun (1942). Skoog (1944) commenced his work on organogenesis in tobacco callus. For rapid propagation, hundreds and thousands of the plants can be raised from small amount of callus cultured in a continuous process. However, one of the serious drawbacks with this method is the production of genetically-unstable plantlets. Hence, Ball (1946), who is considered as the Father of Micropropagation, succeeded in the development and raising of transplantable whole plants (plantlets) of

Lupinus and *Tropaeolum* from the sterile cultures by culturing their shoot tips with a couple of leaf primordia. La Rue (1947) who initiated the endosperm cultures of *Zea mays* obtained subcultures of maize. Skoog and Tsui (1948) studied the chemical control of growth, bud formation in tobacco stem segments and callus cultured *in vitro.* They suggested that callus induction and shoot initiation can be regulated by making some manipulations in the composition of the culture medium. They had demonstrated that the addition of adenine and high levels of phosphate in tobacco pith tissue culture increased the callus growth and bud formation even in the presence of IAA. Street and Dormer (1949) initiated their work on root culture and accounted for its nutrition requirement. The difficulty of culturing the monocots, once considered to be recalcitrant to the conditions of the culture, was overcome by the work of Morel and Wetmore (1951) who were successful in culturing the monocots.

Skoog (1944), and Skoog and Tsui (1948) had demonstrated that the addition of adenine and high levels of phosphate increased the callus growth and bud formation even in the presence of IAA. Cell division in tissue had to await the discovery of KN. This was first discovered by autoclaving freshly-isolated slurries of DNA from herring sperm (Miller *et al.,* 1955). The related analogue, BAP, was then synthesized and it too stimulated cell division in the cultured tissue. Later on, other naturally-occurring plant hormone zeatin was detected in young maize endosperm by Miller (1961), which was isolated by Letham (1963). The discovery of cytokinins has given much impetus to the tissue culture.

For rapid propagation, hundreds and thousands of plants can be raised from small amount of callus cultured in a continuous process. However, one of the serious drawbacks with this method is the production of genetically-unstable plantlets. Hence, Ball (1946) successfully raised transplantable whole plants of *Lupinus* and *Tropaeolum* by culturing their shoot tips with a couple of leaf primordia. However, the demonstration of the practical usefulness of this important technique must be credited to Morel, who with Martin (Morel and Martin, 1952), for the first time recovered virus-free *Dahlia* plants from infected individuals by excising and culturing their shoot tips *in vitro*.

While applying the technique of shoot tip culture for raising virus-free individuals of an orchid, Morel (1960) also realized the potential of this method for the rapid propagation of other plants. This contribution of Morel not only revolutionized the orchid industry, but also gave impetus to the utilization of shoot bud culture for rapid cloning of other plants.

Muir (1953) demonstrated that by transferring callus tissues of *Tagetes erecta* and *Nicotiana tabacum* to liquid medium and agitating the cultures on a shaking machine, it was possible to break the tissue into single cells and small cell aggregates. Later on, Muir *et al.* (1954) also succeeded in mechanically-picking single cells from these shake cultures (suspension cultures) as well as soft callus tissues, and making them divide by individually placing them on separate filter papers resting on the top of a well-established callus culture. Jones *et al.* (1960) designed a microculture method for growing single cells in hanging drops in conditioned medium and Bergmann (1960) designed agar plate method. Vasil and Hildebrandt (1965) raised whole plants starting from single cells of tobacco by filtering suspension cultures and growing the isolated cells on solid media containing 0.6% agar. These techniques are now widely used for cloning tissues. A technology was developed, which provided a source of plant cells in suspension culture that were ideally suited for biochemical and nutritional studies (Gamborg, 1966; Gamborg *et al.,* 1968; Gamborg and Shyluk, 1970).

Cell division in tissue had to await the discovery of KN. This was first discovered by autoclaving freshly-isolated slurries of DNA from Herring sperm by Miller, Skoog, Von Saltza and Strong (1955) who identified a cell division factor, KN (6-furfural aminopurine). The breakthrough, which has made tissue culture applicable to the propagation of diverse plants, has been the discovery by Skoog and Miller (1957) i.e., manipulating the relative concentrations of cytokinin and auxin could regenerate plants. Skoog and Miller advanced a hypothesis that root initiation and shoot initiation in the cultured callus can be regulated by specific ratios of auxins and cytokinins. In the same year, they also discovered and introduced the idea of synergistic action of auxins and cytokinins in promoting cell division in tobacco. In 1958, Reinert (1959) and Steward (1958), for the first time, reported on the production of somatic embryos developing from cells of carrots grown in culture. Somatic embryogenesis has also become a very important system for plant regeneration (Attree and Fowke, 1993). The approach is efficient and there is less chance for genetic abnormalities in the regenerated plants. The technology was greatly enhanced by the use of the medium developed by MS medium (1962), which has become the standard for plant regeneration investigations. The related analogue of KN, BAP was synthesized and noticed that it too stimulated cell division in the cultured tissue. Later, other naturally-occurring plant hormone, zeatin, was detected in young maize endosperm by Miller (1961), which was isolated by Letham (1963). The discovery of cytokinins gave much impetus to the tissue culture. In the year 1965, Vasil and Hildebrandt described rearing of a mature tobacco plant from a single cell grown in microculture. A technology was developed, which provided a source of plant cells in suspension culture, which were ideally suited for biochemical and nutritional studies (Gamborg, 1966; Gamborg *et al.,* 1968; Gamborg and Shyluk, 1970). In 1966, Torrey advanced a hypothesis that organogenesis in callus is initiated with the formation of clusters of meristematic cells (meristemoids). In the same year, Stroun, Anker, Charles and Le Doux succeeded in transferring the DNA in tomato under *in vitro* conditions.

One special aspect of plant regeneration is through anther culture and more recently through microspore culture. Guha and Maheshwari (1964) cultured anthers of *Datura innoxia* on a BM supplemented with CM, KN or grape juice. Numerous embryoids developed inside the anther, which grew into plantlets. Guha and Maheshwari (1966) confirmed that these plantlets are haploid. Haploids are very useful in plant breeding as they can be diploidized by treating with colchicine, and homozygous diploids for all the genes can be obtained. This technique has been used in a large number of species to obtain haploids from anthers and pollen grains.

The Pfizer Company made the first attempt for the *in vitro* industrial production of secondary metabolites during 1950–1960, and Routien and Nickell obtained the first patent in 1956. Brodelius *et al.* (1979) developed the technique of immobilization of plant cells so that the biomass could be utilized for longer periods, besides its other advantages. Mitsui Petrochemical Co. of Japan commercialized the tissue culture product Shikonin from cell cultures of *Lithospermum erythrorhizon* (Curtin, 1983).

Regeneration of the plants from carrot cells frozen at the temperature (-196°C) of LN was first reported by Nag and Street (1973). Thorpe (1980) has made a very valuable contribution towards understanding of the biochemistry and physiology underlying *de novo in vitro* shoot initiation.

The next step in technology development was the research now appearing on producing protoplast from plant cells. Although the isolation of protoplast (Klercker, 1892) and their fusion (Kuster, 1909) were reported almost 100 y ago, it was to the credit of Cocking (1960) whose work introduced the concept of enzymatic isolation of plant protoplast. The isolated protoplasts from tomato fruit locule tissues readily regenerate a new wall (Pojnar *et al.,* 1967). However, it was not until nearly 10 y that the division of isolated tobacco mesophyll protoplasts was reported by Takebe *et al.* (1968). This led the successful culture, following cell wall resynthesis, and the reconstituted cell is capable of sustained divisions (Kao *et al.,* 1970; Nagata and Takebe, 1970). The totipotency of isolated protoplast was demonstrated by Nagata and Takebe (1971) and Takebe *et al.* (1971). Higher plant protoplast can be induced to fuse by a variety of chemical treatments to form intrageneric and intergeneric somatic hybrids (Power *et al.,* 1970; Keller and Melchers, 1973; Kao and Michayluk, 1974; Wallin *et al.,* 1974). Carlson *et al.* (1972) produced first somatic hybrid between *Nicotiana glauca* and *N. langsdorffi* whereas Melchers *et al.* (1978) produced intergeneric hybrid between sexually incompatible parents, potato and tomato.

Larkin and Scowcroft (1981) have proposed a general term 'Somaclonal variation' to describe the genetic variations in regenerated plants from any type of the cell culture. The potential use of somaclonal variation in crop improvement was first demonstrated in sugarcane (Heinz *et al.,* 1977).

Tissue culture headed towards a new research area with the advent of restriction enzymes in early 1970s. The totipotent plant cells could now be altered by insertion of specific foreign genes giving rise to genetically-modified crops. In 1970, Smith and Nathans isolated first restriction enzyme from *Haemophilus influenzae,* which was later purified and named *Hind III.* Same year witnessed other Nobel prize winning discovery by Baltimore who isolated reverse transcriptase from RNA tumour viruses. This is a useful enzyme in genetic engineering, which functions to convert RNA to DNA, and hence useful in construction of cDNA from mRNA. Another path-breaking discovery establishing potential of genetic engineering came in 1972 when Paul Berg working at Stanford University produced first recombinant DNA *in vitro* by combining DNA from SV40 virus with that of lambda virus. This led to construction of first recombinant organism by Cohen and Boyer in 1973. Genetic engineering's potential was first exploited when a man-made insulin gene was used to manufacture a human protein in bacteria.

Agrobacterium tumefaciens plays a crucial role in plant genetic engineering. The involvement of this bacterium in crown gall disease in plants was recognized as early as 1907 by Smith and Townsend. However, it was in 1974, that Zaenen *et al.* discovered that Ti plasmid is the tumour-inducing principle of *Agrobacterium.* This was followed by its successful integration in plants by Chilton *et al.* in 1977. Zambryski *et al.* in 1980 isolated and studied the detailed structure of T-DNA and its border sequences. Soon thereafter in 1984, transformation of tobacco with *Agrobacterium* was accomplished to develop transgenic plants. Simultaneously, there was an upsurge in the development of the techniques of genetic engineering in mid 1970s. Sanger *et al.* (1977), and Maxam and Gilbert (1977) reported the techniques for large-scale DNA sequencing. This was followed by complete genome sequencing projects on many prokaryotes and eukaryotes such as *Haemophilus influenzae* in 1995, *E. coli* in 1997. Human genome was sequenced successfully in 2001, thus laying the foundation of genomics, which is the main focus of present- day biotechnology.

The current emphasis and importance of plant biotechnology can be gleaned from the IXth International Congress on Plant Tissue and Cell Culture held in Israel in June, 1998. The theme of the Congress was 'Plant Biotechnology and *In vitro* Biology in the 21st Century'. This theme was developed through a scientific program, which focused on the most important developments, both basic and applied, in the areas of plant tissue culture and molecular biology, and their impact on plant improvement and biotechnology (Thorpe and Lorz, 1998). The titles of the plenary lectures were (1) 'Plant Biotechnology Achievements and Opportunities at the Threshold of the 21st Century', (2) 'Towards Sustainable Crops via International Cooperation', (3) 'Signal Pathways in Plant Disease Resistance', (4) 'Pharmaceutical Foodstuffs: Oral Immunization with Transgenic Plants', (5) 'Plant Biotechnology and Gene Manipulation', and (6) 'Use of Plant Roots for Environmental Remediation and Chemical Manufacturing'. These titles not only clearly show where tissue culture was, but also show where it was heading, as an equal partner with molecular biology as a tool in basic plant biology and in various areas of application. Later Congresses in Florida, USA (2002), Beijing, China (2006), and St Louis, Missouri, USA (2010) supported this view and clearly demonstrated the advances that were being made in these areas. As Schell (1995) pointed out, progress in applied plant biotechnology is fully matching and is, in fact, stimulating fundamental scientific progress.

Tissue culture is, thus, proving to be useful in a variety of ways including plant propagation, raising and maintenance of high-health status plants, germplasm storage, and a valuable technique in plant improvement. In plant improvement, tissue culture may be used in the technique of gene insertion to improve plant stocks. Plant tissue culture has also been used in the production of secondary metabolites in plants. The production of flavours, sweeteners, natural colourants as well as pharmaceuticals was achieved using tissue culture.

Table 1: Landmarks in plant tissue culture research

Worker	Year	Contribution
G. Haberlandt	1902	First to culture isolated plant cells *in vitro* on artificial medium
Hannig	1904	Embryo culture of selected crucifers
W.J. Robbins and W. Kotte	1922	Culture of isolated roots – organ culture
F. Laibach	1925	Use of embryo culture technique in interspecific crosses of *Linum*
P.R. White	1934	Demonstrated capability of indefinite culture of tomato roots (long periods)
R.J. Gautheret and P. Nobecourt	1939	First culture of cambial tissues isolated from carrot
P.R. White	1939	Callus culture of tobacco tumour tissue from interspecific hybrid *Nicotiana glauca* X *N. langsdorfii*
J. Van Overbeek	1941	Discovered nutritional value of liquid endosperm of coconut (CM) for culture
P.R. White and A.C. Braun	1942	Experiments on crown gall and tumour formation in plant growth of bacteria-free crown gall tissue
F. Skoog	1944	*In vitro* adventitious shoot formation in tobacco
A. Caplan and F. C. Steward	1948	Use of CM and 2,4-D for proliferation of cultured carrot and potato tissues
G. Morel	1950	Culture of monocot tissues using CM
E.A. Ball	1950	Regeneration of organs from callus tissue of *Sequoia sempervirens*
G. Morel and C. Martin	1952	Use of meristem culture to obtain virus-free Dahlias

Worker	Year	Contribution
W.H.Muir	1953	Developed technique for culture of single isolated cells (nurse culture method)
W. Tulecke	1953	Production of haploid callus of the gymnosperm *Ginkgo biloba*
C.O. Miller and F. Skoog	1955	Discovery of cytokinins (KN) as potent cell division factor
F. Skoog	1957	Predicted hypothesis that shoot and root initiation in cultured cells is regulated by the proportion of auxin and cytokinin in the culture medium
P.Maheshwari and Rangaswamy	1958	Regeneration of somatic embryos *in vitro* from the nucellus of *Citrus*
J. Reinert and F.C. Steward	1959	Regeneration of embryos from callus clumps
E.C.Cocking	1960	Enzymatic isolation and culture of protoplasts
G. Morel	1960	Use of modified shoot apex technique for orchid propagation
T. Murashige and F. Skoog	1962	Development of MS medium
S.Guha and S.C. Maheshwari	1966	Cultured anthers can produce haploid plants
Carlson	1970	Selection of biochemical mutants *in vitro* by the use of tissue culture
Carlson *et al.*	1972	Interspecific hybridization through protoplast fusion
Berg	1972	Produced first recombinant DNA combining SV40 virus and λ virus
Zaenen *et al.*	1974	Discovered Ti plasmid as tumour-inducing principle of *Agrobacterium*
J.P. Nitsch	1974	Generated plants from cryopreserved plant tissues
O'Farrel	1975	Developed high-resolution two-dimensional gel electrophoresis system
Bajaj *et al.*	1976	Production of somatic embryos by using protoplast fusion technique
Chilton *et al.*	1977	Successfully integrated Ti plasmid DNA from *Agrobacterium tumefaciens* in plants
Sanger, Maxam-Gilbert	1977	Gave technologies for DNA sequencing

Worker	Year	Contribution
G.Melcher	1978	Discovered the chromosomal doubling in haploid tissues in the medium
Marton *et al.*	1979	Co-cultivation procedure developed for transformation of plant protoplasts with *Agrobacterium*
Alfermann *et al.*	1980	Use of immobilized whole cells for biotransformation of digitoxin into digoxin
Zambryski	1980	Detailed structure of T-DNA and border sequences
Siderov *et al.*	1981	Introduction of the term 'somaclonal variation'
Larkin and Scowcroft	1981	Isolation of auxotrophs by large-scale screening of cell colonies derived from haploid protoplasts of *Nicotiana plumbaginifolia* treated with mutagens
Krens *et al.*	1982	Protoplasts are able to incorporate naked DNA; transformation with isolated DNA is consequently possible.
Zimmermann	1982	Fusion of protoplasts by electrical stimulus
Pelletier *et al.*	1983	Intergeneric cytoplasmic hybridization in radish and rape
Kary Mullis	1983	Invented PCR for amplification of DNA.
Horsh *et al.*	1984	Developed transgenic tobacco by transformation with *Agrobacterium*
Paszkowski *et al.*	1984	Transformation of plant cells with plasmid DNA
Alec Jeffreys	1984	Development of the genetic fingerprinting technique for identifying individuals by analyzing polymorphism at DNA sequence level
Horsch *et al.*	1985	Developed using cDNA of coat protein gene of TMV
Powell-Abel *et al.*	1986	TMV-resistant tobacco and tomato transgenic plants
Sanford *et al.*; Klein *et al.*	1987	Development of biolistic gene transfer method for plant transformation
Barton *et al.*	1987	Isolation of Bt gene for bacterium (*Bacillus thuringiensis)*
Klien *et al.*	1987	Developed biolistic gene transfer method for plant transformation
Fleischmann *et al.*	1995	Sequenced *Haemophilus influenzae*

Worker	Year	Contribution
Blattner *et al.*	1997	Sequenced *E. coli* genome
Venter *et al.*	2001	Human Genome Project consortium and sequenced human genome successfully
	2005	Rice genome sequenced under International Rice Genome Sequencing Project

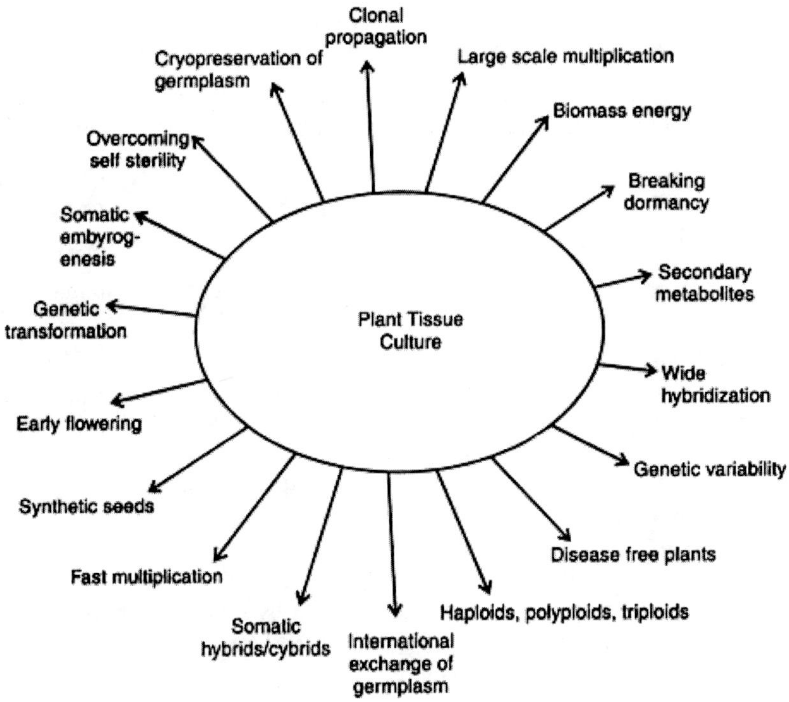

Fig. 3 Different applications of plant tissue culture

1.5 APPLICATIONS OF PLANT TISSUE CULTURE

Plant tissue culture now has direct commercial applications as well as value in basic research into cell biology, genetics and biochemistry. The techniques include culture of cells, anthers, ovules and embryos on experimental to industrial scales, protoplast isolation and fusion, cell selection, and meristem and bud culture. The applications include (Fig. 3).

- Micropropagation using meristem and shoot culture to produce large numbers of identical individuals
- Screening programmes of cells, rather than plants, for advantageous characters
- Large-scale growth of the plant cells in liquid culture as a source of secondary products
- Crossing distantly-related species by protoplast fusion and regeneration of the novel hybrid
- Production of dihaploid plants from haploid cultures to achieve homozygous lines more rapidly in breeding programmes
- As a tissue for transformation, followed by either short-term testing of genetic constructs or regeneration of transgenic plants
- Removal of viruses by propagation from meristematic tissues

1.6 TERMINOLOGY

- **Agar or agar-agar:** A gelatinous polysaccharide obtained from the red alga *Gelidium corneum* and from several other red algae. It is a solidifying agent that is mixed with nutrient media (0.6–1%), and it forms a gel for growing tissue culture plants and for other puroposes. It ranges in quality from relatively inert to very impure (complex, undefined). It's firmness as gelling agent is affected by the medium pH (softer when the medium is more dilute). Agar gels melt at about 100°C, but solidify at about 44°C.
- **Aliquot or aliquot part:** An evenly-divided unit, portion or sample (fraction) of the whole.
- **Androgenesis:** Development of haploid plants from the male gametophyte following a developmental pattern resembling embryogenesis, resulting from the culture of anthers or microspores
- **Asepsis:** Without infection or contaminating microorganisms
- **Aseptic:** Free of pathogens, contaminants, algae, bacteria, fungi, viruses, etc.

- **Auxotroph:** A mutant organism that will not grow on a minimal medium, but requires the addition of some growth factor.
- **Axenic:** A pure culture of single organism that is not contaminated by symbionts or parasites.
- **Basal medium:** A medium containing inorganic salts of major and minor elements, vitamins (without hormones) and sucrose is known as BM.
- **Batch culture:** A type of cell suspension culture in which cell or small cell aggregates grow in a definite volume of nutrient medium.
- **Bioreactor:** A vessel used for large-scale aseptic culture of cells, embryos or shoots under controlled growth conditions.
- **Callus:** An unorganised mass of parenchymatous tissue formed in response to wounding of plant parts or in tissue culture is called as callus (plural: calli).
- **Clonal propagation:** Asexual reproduction or vegetative propagation of plants considered being physiologically and/ or genetically uniform and which originated from a single individual or explant.
- **Cybrid:** A cytoplasmically-hybrid cell with organelles from both parental sources e.g., achieved by fusion of a cytoplast with a whole cell.
- **Cytodifferentiation:** Cell differentiation is the basic event of development in higher organisms and conveniently referred to as cytodifferentiation.
- **Dedifferentiation:** The phenomenon of a mature cell reverting to its meristematic state and forming undifferentiated callus tissue is termed as dedifferentiation.
- **Embryoid:** Mass of cells, which has an external morphology resembling a proembryo, non-zygotic embryo developed in tissue culture.
- **Embryo rescue:** Application of embryo culture technique to obtain interspecific hybrid plants prior to abortion of the embryos due to post-fertilization hybridization barriers.

- **Excise:** Removing the part of a plant or seed in aseptic conditions for tissue culture
- **Explant:** The tissue taken from a plant or seed and transferred to culture medium to establish a tissue culture system or regenerate a plant
- ***Ex vitro:*** (Latin: from glass) Organisms removed from culture and transplanted generally to soil or potting mixture
- **Friability:** The property of cultured plant cells that separate easily from one another
- **Gynogenesis:** Development of haploid plants from the female gametophyte by culture of unfertilized ovules or ovaries
- **HEPA filter:** A filter capable of screening out particles larger than 0.3 microns. They are used in laminar airflow cabinets (hoods) for sterile transfer work.
- ***In vitro*:** (Latin = in glass) Experimentation on organisms or portions thereof in glassware or culture; growing under artificial conditions as in tissue culture. The antonym is *in vivo*.
- **Micropropagation:** Synonymous with *in vitro* propagation especially with reference to enhanced axillary branching or adventitious plantlet regeneration.
- **Morphogenesis:** The process of growth and development of a form or structure.
- **Organogenesis:** The initiation (*de novo*) and growth of the organs (roots and shoots usually) from cells or tissues, as in organ culture. Organs may form on the surface of explants (direct organogenesis) or upon an intervening callus phase (indirect organogenesis).
- **Packed cell volume:** A quantitative method of estimating cell growth. It is based on the total cell volume in an aliquot of suspension culture. The aliquot is centrifuged for 5 min at 200 g after which the PCV is expressed as a percentage of the aliquot volume.
- **Passage:** The transfer of cell or tissue from old culture medium to fresh medium within a definite time period. Maintenance of the culture in a medium for a definite period

before transferring to fresh medium constitutes a passage. This term is synonymous with the term 'subculture'.

- **Plating Efficiency:** It is a quantitative assessment of the percentage of cell colonies per total number of cells or protoplasts plated at a definite density for a specific time period. It can be determined by counting cell colonies under microscope at the end of experiment and calculated by using the formula

$$PE = \frac{\text{No. of colonies/plate}}{\text{No. of cell units}} \times 100$$

- **Polarity:** An interesting observation made by Vochting (1878) suggested 'polarity' as a characteristic feature guiding the development of plant fragments. In his classical experiments on stem cuttings, Vochting observed that the upper portion of a piece of stem always produced buds, and the basal region produced callus or roots.

- **Protocol:** A sequence of activities, techniques or procedures linked to achieve a specific goal; as a schedule for micropropagation.

- **Redifferentiation:** The ability of a dedifferentiated cell to form a whole plant or plant organs is termed as redifferentiation.

- **Somaclonal variation:** Variation, which occurs in cultures of the cells and tissues that may be either genetic or epigenetic.

- **Synthetic seeds:** Somatic embryos encapsulated in hydrated or desiccated coating, which protects the embryo from mechanical damage during handling, and allows their germination such as sexual seeds. The capsule may also serve as an artificial endosperm.

- **Totipotency:** The ability of a plant cell to develop into a complete plant is known as totipotency.

- **Vitrification:** A physical process by which a concentrated aqueous solution cooled to low or superlow temperature directly solidifies into an amorphous glassy state without crystallization; the term was earlier used to hyperhydration.

1.7 BOOKS FOR FURTHER READING

Arditti J 2008 *Micropropagation of Orchids*, New York: Agritech Publications

Bhojwani SS 1990 *Plant Tissue Culture: Applications and Limitations*, New York: Elsevier Science Publications

Bhojwani SS and Razdan MK 1983 *Plant Tissue Culture: Theory & Practice*, Amsterdam: Elsevier Scientific Publications

Chawla HS 2002 *Introduction to Plant Biotechnology,* New Delhi: Oxford & IBH

Collins HA and Edwards S 1998 *Plant Cell Culture*, Oxford: Bios Scientific Publishers

De KK 1992 *An Introduction to Plant Tissue Culture*, Calcutta: New Central Book Agency

Debergh PC and Zimmerman RH (eds) 1991 *Micropropagation, Technology and Application,* Dordrecht: Kluwer Academic Publishers

George EF, Hall MA, and De Klerk G 2008 *Plant Propagation by Tissue Culture,* Edington, England : Exegetics Ltd.

Johri BM 1982 Experimental Embryology of Vascular Plants, New York: Springer Verlag

Kartha KK 1985 Cryopreservation of Plant Cell and Organs, Boca Raton: CRC Press

Kyle L and Kleyn J 1997 *Plants from Test Tubes,* Portland: Timber Press

Purohit SS 2003 *Plant Tissue Culture,* Jodhpur: Student Edition

Raghavan V 1976 Experimental Embryogenesis in Vascular Plants, London: Academic Press

Raghavan V 1986 Embryogenesis in Angiosperms: A developmental and experimental study, New York: Cambridge Univ. Press

Razdan MK 2002 An Introduction to Plant Tissue Culture, New Delhi: Oxford & IBH

Vasil IK and Thorpe TA 1994 *Plant Cell and Tissue Culture,* The Netherlands: Kluwer Academic Publishers

White PR 1963 The Cultivation of Animal and Plant Cell, New York: Ronald Press.

■ ■ ■ ■

2 LABORATORY REQUIREMENTS FOR PLANT TISSUE CULTURE

2.1 LABORATORY ORGANIZATION

Any laboratory, in which tissue culture techniques are performed regardless of the specific purpose, must contain a number of basic facilities. These usually include the following:

- A general washing area
- A media preparation, sterilization and storage area
- An aseptic transfer area
- Environmentally-controlled incubators or culture rooms
- An observation/data collection area

2.1.1 Media Preparation Room

Media preparation room should have storage and bench space for keeping the glassware, chemicals and minor equipments.

- **pH meter (digital):** 230 v 50 Hz, single phase supply with combined pH electrode, range 0–14, temperature compensation 0–100°C. It is necessary for the measurement and adjustment of pH of the nutrient medium.
- **Weighing balances:** Three types of weighing balances *viz.*, pan balance, chemical balance and electric balance are required for weighing the chemicals, sugars, agar-agar, callus (fresh and dry weight) and others. Manual – single pan, capacity 100–200 g and sensitivity 0.1 mg; electronic top-loading balance – precision 0.005 g, range 0–120 g; semimicrobalance – range 0.00001–120 g.

- **Electric hot air oven:** Thermostatically controlled, range 5–250°C. It is necessary for drying the washed glassware. For this purpose, a number of enameled trays of different sizes are required for keeping wet glassware inside the oven.
- **Microwave oven:** For heating the medium to dissolve the agar during medium preparation
- **Laminar airflow cabinet:** One type of transfer area found convenient and currently used in most of the laboratories is a laminar airflow cabinet (Fig. 4). A small motor blows the air into the unit first through a coarse filter, where large particles are separated, and subsequently through a 0.3 μm HEPA filter. The air is directed either downward (vertical flow unit) or outward (horizontal flow unit) over the working surface. The air coming out of the fine filter is ultra clean (free from fungal or bacterial contaminants) and its velocity (27 ± 3 m/min) adequately prevents the microcontamination of the working area used by a worker sitting in front of the cabinet. Various other contaminants (hair, salts, flakes) are also blown away by the ultraclean airflow. In this way, an aseptic environment is maintained as long as the cabinet is switched on.

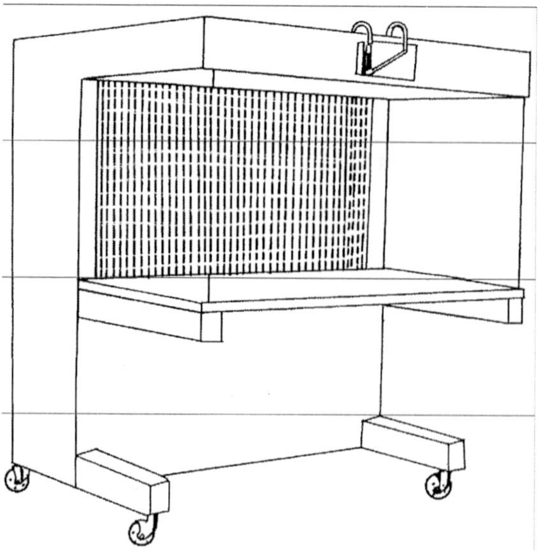

Fig. 4 A laminar flow chamber

- **Microscopes:** Dissecting microscope, laboratory microscope with photomicrographic attachment and inverted microscope are essential for various purposes.
- **Electrical autoclaves:** Autoclaves are used for sterilizing the media. In case autoclaves are not available, pressure cooker can also be used. Vertical or horizontal models with safety valves, pressure gauge and steam-release cock are generally used.
- **Double-distilled water equipment:** It should have all glass (or quartz) water stills, heat-resistant boiling flask and built in energy regulator. Better to have auto power cut-off in case water supply is stopped. A single-distillation and a double-distillation water plant are indispensable. Two big plastic containers are required for storing the distilled water.
- **Gyratory shaker:** Capacity 50–1,000 ml flasks, speed 80–200 rpm. They are useful for suspension culture.
- Refrigerator for safe storage of chemicals and media stocks. It is essential for storing various thermolabile chemicals such as vitamins, hormones, amino acids, CH, yeast extract, CM etc. Stock solutions of salts are also stored in the refrigerator to prevent contamination.
- **Vacuum pump:** It is required for filtering liquid media, sugar solution etc. through filter apparatus using air suction.
- **Heater:** It is needed for heating or warming the medium to dissolve agar or to melt the agarified medium.
- Hot plates
- Filter units
- Vacuum cleaner
- Water deionizer
- Gas stove
- Spirit lamps
- **Glassware:** Different types of glassware and apparatus are used to culture the plant tissues. These glassware and apparatus are both conventional and also specific. The glassware should necessarily be of Pyrex or Corning or Borosilicate glass and

should be autoclavable. The glassware is represented by the measuring cylinders, conical flasks, pipettes, beakers, etc. Plant tissues are generally grown in wide-necked Erlenmeyer conical flasks (100 ml or 150 ml or 200 ml), culture tubes (25 mm in diameter and 150 mm in length), petri plates (50 mm or 90 mm or 140 mm in diameter) and screw-capped universal bottles (of 20 cm^3 capacity). Sometimes, jam bottles or milk bottles may also be used. Proper care must be taken to ensure that the glassware is properly cleaned (either new or dirty) before use. The traditional way of cleaning the glassware is to soak them in soap water for some time, followed by brushing and washing them well with tap water, and finally rinsing them with one-time distilled water. Later, they are dried in a hot air oven and then stored in a dust-proof cupboard or a drawer. Prior to autoclaving, all the culture vessels (flasks and tubes) must be fitted with cotton plugs (tightly rolled plugs of non-absorbent cotton wrapped in gauge) to allow free gaseous exchange, but not to allow the entry of microbes. The exposed parts of the plug as well as the rim of the culture vessel should be covered by a brown paper or an aluminium foil, to keep free from dust and also protect the plug not to become wet during autoclaving. Sometimes, presterilized, disposable, autoclavable, plasticware is used to culture the plant tissues. In case of sterilization of the medium containing thermolabile compounds or enzymes (for protoplast isolation), a specially-designed glass made of bacterial filter or an autoclavable plastic-made bacterial filter is used. A small spirit lamp made of glass is essential for the flame sterilization of the instruments using methylated spirits.

- **Chemicals:** AR grade
- Detergents
- Test tube/culture vial stands
- Laboratory work benches

The water used in preparing media must be of the utmost purity and highest quality. Tap water is unsuitable because it may contain cations (ammonium, calcium, iron, magnesium, sodium, etc.), anions (bicarbonates, chlorides, fluorides, phosphates, etc.), microorganisms (algae, fungi, bacteria), gases (oxygen, carbon dioxide, nitrogen) and particulate matter (silt, oils, organic matter, etc). Water used for plant tissue culture should meet, at a minimum, the standards for type II reagent grade water i.e., be free of pyrogens, gases and organic matter, and have an electrical conductivity less than 1.0 μmho/cm.

The most common and preferred method of purifying water to type II standards is a deionization treatment followed by one or two glass distillations. The deionization treatment removes most ionic impurities, and the distillation process removes large organic molecules, microorganisms, and pyrogens. Three other methods that will produce type II purity water are absorption filtration, which uses activated carbon to remove organic contaminants and free chlorine; membrane filtration, which removes particulate matter and most bacterial contamination; and RO, which removes approximately 9% of the bacterial, organic and particulate matter as well as about 90% of the ionized impurities. The most universally reliable method of water purification for tissue culture use is a deionization treatment followed by one or two glass distillations, although simple deionization alone is sometimes successfully used. In some cases, newer RO purifying equipment (Milli-RO™, Millipore™, RO pure™, Barnstead™, Pierce™) combined with cartridge ion exchange, adsorption, and membrane filtering equipment has replaced the traditional glass distillation of water.

2.1.2. Washing Area

This is very important for tissue culture laboratory. The washing area should contain large sinks, some lead-lined to resist acids and alkalis, draining boards and racks. It should also have an access to demineralized water, distilled water, and double-distilled water. Space for drying ovens or racks, automated dishwashers, acid baths, pipette washers and driers, and storage cabinets should also be available in the washing area. It should be provided with running hot and cold tap water, brushes of various sizes, detergent and a bucket of single-distilled water for a final rinse of the washed glassware.

2.1.3. Transfer Area

Tissue culture techniques can be successfully performed on an open laboratory bench under very clean and dry atmospheric conditions. However, it is advisable that a laminar flow hood or sterile transfer room be utilized for making transfers. There should be a source of electricity, gas, compressed air and vacuum within the transfer area.

The most desirable arrangement is a small dust-free room equipped with an overhead UV and positive-pressure ventilation unit. The ventilation should be equipped with a HEPA filter. A 0.3 μm HEPA filter of 99.97–99.99% efficiency works well. All surfaces in the room should be designed and constructed in such a manner that the dust and microorganisms do not accumulate, and the surfaces can be thoroughly cleaned and disinfected. A room of such design is particularly useful if large numbers of cultures are being manipulated or large pieces of equipment are being utilized.

Another type of transfer area found convenient and currently used in most of the laboratories is a laminar airflow cabinet, details of which are given earlier.

Ordinarily, a small wooden hood (Fig. 5) may also be used for inoculation work. The hood has a glass or plastic door either sliding or hinged. The door should be dust proof. The culture hoods can be conveniently placed on a bench in a quiet corner of the laboratory. The worker inserts his hands through two holes in front of the chamber to carry out the operations. The transparent glass or plastic doors enable the worker to see inside and make the necessary manipulations.

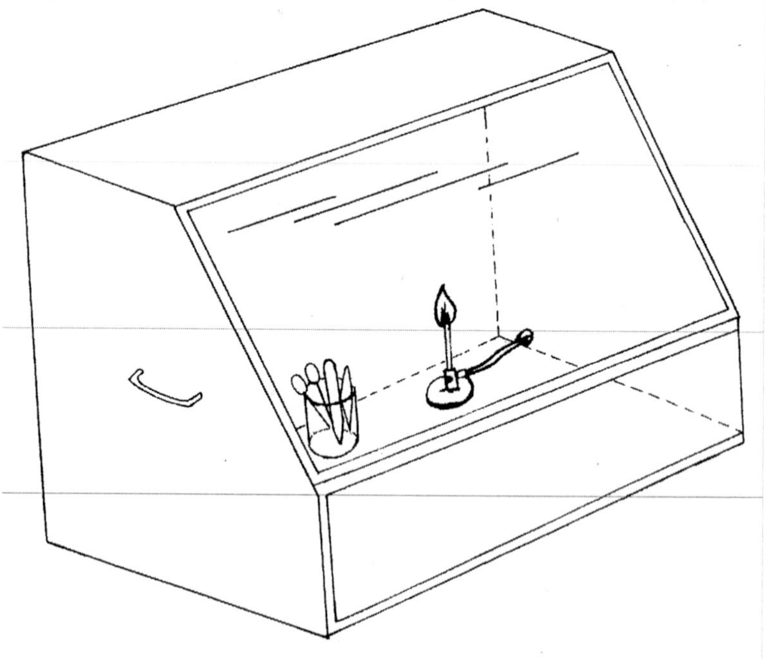

Fig. 5 Design for a simple inoculating hood

2.1.4. Instruments

The routinely-used instruments for culture work include various-sized scalpels and foreceps, spatulas, scissors, etc. (Fig. 6). All the instruments should be of stainless steel. A list of commonly-used instruments, their measurement and uses are given below.

Table 2: Commonly-used instruments, their measurement and uses

Instruments	Measurement in length (cms)	Size No.	Use
1. Scalpel	20.32, 25.4 etc.	-	For cutting the plant material or explants
2. Surgical scalpel			
(a) Holder	--	3, 4	For cutting the plant material or explants
(b) Disposable blade (arrow headed)	--	11	
3. Forceps (without corrugated marking at the pointed end)	20.32, 25.5	--	For holding the material
4. Jewellery fine foreceps	--	5	For peeling leaf epidermis
5. Arrow-headed sharp needle	25.4	--	For dissecting out anthers
6. Spatula (one end spoon-headed and opposite end flat)	25.4	--	For transferring of the material
7. Metal cork borers (1 cm diameter)	--	--	For making tissue cylinder e.g., potato, carrot etc.
8. Scissors	25.4	--	For cutting roots or shoots from aseptic culture

2.1.5. Culture Room

Plant cultures are greatly influenced by physical factors such as temperature, relative humidity and light. A constant temperature of 25 ± 2°C is maintained in culture room by the air conditioner. All types of plant tissues are incubated under conditions of well-controlled temperature, humidity, illumination and air circulation. Specially-designed shelves made of glass, rigid wiremesh or wood are provided in the culture room for storing cultures. Each shelf is illuminated by cool white fluorescent tubes with 2,000 lux light. Usually, two tubes are provided for each shelf. A photoperiod of 16 h light and 8 h dark is maintained with an automatic timer. The cultures are provided with a light intensity of 10 W/cm^2. A thermometer and a hygrometer are fixed on the wall at the safety corner of the room to check the temperature and relative humidity respectively. The relative humidity of the culture room is maintained above 50%.

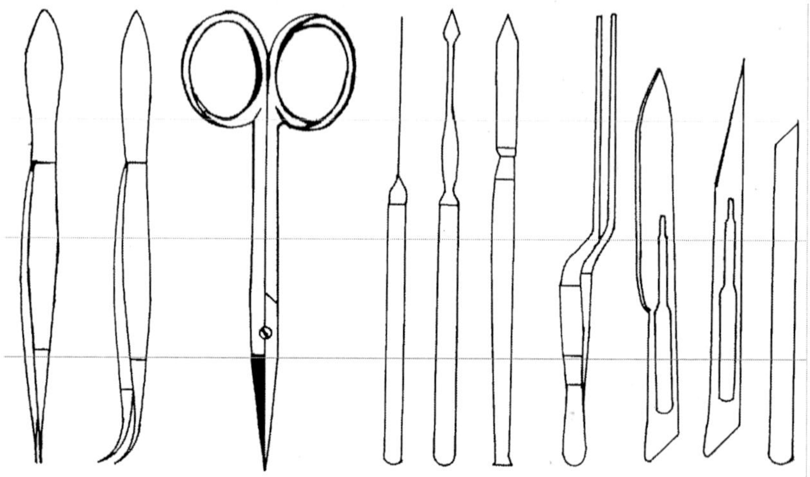

Fig. 6 A set of instruments used for tissue culture work

The culture room should also have a shaker for suspension culture or single cell culture in moving liquid medium. The speed of revolution of the shaker can be controlled. The shaker is also provided with light. The platform of the shaker is fitted with clips for holding the conical flasks (150–200 ml). However, the optimum temperature, light and shaker conditions vary depending upon the plant species being cultured.

2.1.6. Transplantation Area

Plants regenerated from *in vitro* tissue cultures are transplanted to the soil in pots. The potted plants are ultimately transferred to greenhouses or growth cabinets for further observations under controlled conditions of light, temperature and humidity.

Prior to transfer of the potted plants to greenhouse, the acclimatization of these plants may begin in growth chambers under mist benches, or in standard greenhouse bay. Even if funds are a deterrent for providing greenhouse facilities, it is, however, essential that minimum facility maintaining high humidity by fogging, misting, or a fan system for cooling in summer and heating system for winter, are provided in a potting room adjacent to the tissue culture laboratory.

2.1.7. Laboratory and Personal Safety

Every new worker should be given some orientation before starting the work. It should include instructions on operation of various equipments such as pH meter, balances, laminar airflow chamber, microscope, gas burners, etc. Special instructions should also be given on the potential for controlling fire, broken glass, chemical spills or accidents with sharp-edged instruments.

■■■■

3 TISSUE CULTURE MEDIA

3.1. INTRODUCTION

In vivo plant cells, tissues and organs get their appropriate nutrients and growth requirements from the intact plant body for their organized growth and development. Isolated cells, tissues and organs also need nutrients for their *in vitro* growth and development. Hence, nutrients are supplied artificially according to the medium formulated by several workers.

3.2. MEDIA COMPOSITION

A number of media have been devised for specific tissues and organs. White's (1953) medium is one of the earliest plant tissue culture media. Other media include MS (1962) medium, B_5 medium (Gamborg *et al.,* 1968) and Nitsch's medium (Nitsch and Nitsch, 1969). The WM is originally formulated for root culture. To induce organogenesis and regeneration of plants in cultured tissues, MS and LS media are widely used. The B_5 medium, originally designed for cell suspension or callus cultures, has (with modifications) proved valuable for protoplast culture. Another medium, designated as N_6, has been developed for cereal anther culture and other tissues, though the medium devised by Nitsch and Nitsch (1969) is frequently used in the experiments on anther culture. Most of these media contain inorganic salts of major and minor elements, vitamins and sucrose (Table 3). A medium with these components is known as BM. Growth regulators such as auxins, cytokinins and gibberellins may be added to this medium. Natural plant extracts such as CM, yeast extract, CH and various fruit juices are also added to the medium for promoting growth in some tissues.

Table 3: Nutritional components of some plant tissue culture media[1]

Components	Amount (mg/l)			
	White's	**MS**	**B_5**	**Nitsch's**
Macronutrients				
Magnesium sulphate ($MgSO_4$ $7H_2O$)	750	370	250	185
Potassium phosphate (KH_2PO_4)	--	170	--	68
Sodium phosphate ($NaH_2PO_4.H_2O$)	19	--	150	--
Potassium nitrate (KNO_3)	80	1900	2500	950
Ammonium nitrate (NH_4NO_3)	--	1650	--	720
Calcium chloride ($CaCl_2.2H_2O$)	--	440	150	--
Ammonium sulphate ($(NH_4)_2SO_4$)	--	--	134	--
Micronutrients				
Boric acid (H_3BO_3)	1.5	6.2	3	--
Manganese sulphate ($MnSO_4.$ $4H_2O$)	5	22.3	--	25
Manganese sulphate ($MnSO_4.$ H_2O)	--	--	10	--
Zinc sulphate ($ZnSO_4.7H_2O$)	3	8.6	2	10
Sodium molybdate ($Na_2MoO_4.2H_2O$)	--	0.25	0.25	0.25
Copper sulphate ($CuSO_4.5H_2O$)	0.01	0.025	0.025	0.025
Cobalt chloride ($CoCl_2.6H_2O$)	--	0.025	0.025	0.025
Potassium iodide (KI)	--	0.83	0.75	0.75
Ferrous sulphate ($FeSO_4.7H_2O$)	--	27.8	--	27.8
Sodium salt of EDTA ($Na_2EDTA.2H_2O$)	--	37.3	--	37.3
EDTA Na Ferric salt	--	--	43	--
Sucrose (g)	20	30	20	20
Organic supplements				
Vitamins				
Thiamine HCl	0.01	0.5	10	0.5
Pyridoxine HCl	0.01	0.5	1	0.5

Components	Amount (mg/l)			
	White's	MS	B_5	Nitsch's
Nicotinic acid	0.05	0.5	1	5
Myoinositol	--	100	100	100
Others				
Glycine	3	2	--	2
Folic acid	--	--	--	0.5
Biotin	--	--	--	0.05
pH	5.8	5.8	5.5	5.8

[1]Growth regulators and complex nutrient mixtures used by various workers are not included here as the quantity and constituents of these compounds vary for specific tissue and organ

[2] **Abbreviations and references**

1. White's 1953 *Am. J. Bot.* **40** 517–524
2. MS: Murashige and Skoog 1962 *Physiol. Plant.* **15** 473
3. Gamborg *et al.,* 1968 *Exp. Cell Res.* **50** 151
4. Nitsch and Nitsch 1969 *Science N.Y.* **163** 85

3.2.1. Inorganic Nutrients

A variety of mineral elements (salts) supply the needed macronutrients and micronutrients required in the life of a plant. Elements required in concentration greater than 0.5 mM/l are referred to as macronutrients and those less than 0.05 mM/l as micronutrients. When mineral salts are dissolved in water they undergo dissociation and ionization. In a medium, one type of ion may be contributed by more than one salt. For example, nitrate (NO_3) ions are contributed by ammonium nitrate and potassium nitrate in MS medium, and potassium (K^+) ions by Potassium nitrate and potassium phosphate. The inorganic and organic chemicals used in the preparation of the media should be of analytical grade i.e., AR or GR.

The macronutrients include six major elements: nitrogen, phosphorous, potassium, calcium, magnesium and sulphur present as salts that constitute various media. All these are essential for plant cell and tissue growth. Culture media should contain at least 25 m mol/l nitrate and potassium. However, considerably better results are obtained if the source for nitrogen in the media is contributed by both nitrates and ammonium (2–20 m mol/l) or any other reduced nitrogen source. When nitrate and ammonium ions are present together in the culture medium, the latter are used more rapidly.

Nitrogen is one of the main elements contributing to the growth of plants *in vitro* and *in vivo*. It is a constituent of amino acids, proteins, certain hormones and chlorophyll. The source of nitrogen *in vitro* could be either organic or inorganic. An indirect effect of nitrogen on tissue growth is through its influence on the pH of the medium. The form of nitrogen, as ammonium (NH_4+) ions or nitrate ions, has a dramatic influence on the morphogenic response of the plant tissues *in vitro*.

Phosphorus is vital for cell division as well as in storage and transfer of energy in the plants. It is found in meristematic and other fast-growing tissues. It is a part of DNA and ATP molecules, and is an essential element in photosynthesis and respiration. It affects plant maturation and root growth. Stunted growth and reddish to purple colouring (anthocyanin) can be symptoms of phosphorus deficiency. Potassium phosphate and sodium phosphate are routinely included in the tissue culture media.

Potassium is essential for normal cell division and it promotes meristematic growth. It is involved in many reactions within plants; however, it is not an actual component of plant protoplasm, fats or carbohydrates. It helps in synthesizing carbohydrates and proteins, manufacturing chlorophyll, and reducing nitrates. Non-availability of potassium results in weak and abnormal plants, sometimes with mottled, curled or dead leaf edges. Potassium nitrate and potassium phosphate are common sources of potassium in culture media; potassium chloride (KCl) is used occasionally.

Calcium is an integral compound of plant cell walls and found in the form of calcium pectate. Calcium plays a role in the formation of pectin (a substance that bonds cell walls together). Calcium helps control permeability and facilitates the movement of carbohydrates and amino acids throughout the plant; it also promotes root development. It assists in growth and development as well as in nitrogen assimilation. As calcium oxalate, it ties up oxalic acid, which is toxic by-product of protein metabolism. Dead shoot or root tips can be a sign of lack of calcium. Calcium is usually included in tissue culture media as calcium chloride or as calcium nitrate ($Ca[NO_3]_2.4H_2O$); tribasic calcium phosphate ($Ca_{10}[PO_4]_6[OH]_2$) is also sometimes included.

Sulphur is found in some proteins, and promotes root development and keeps green foliage. It is supplied in tissue culture media in sulphate (SO_4^{2-}) compounds of magnesium, manganese, zinc, copper etc.

Magnesium is the central element in chlorophyll molecules and is important as an enzyme activator. Its deficiency causes older leaves to become chlorotic. Most tissue culture formulas include magnesium sulphate, commonly known as Epsom salts.

Inorganic elements are required in small quantity, but are essential for the plant cell and tissue growth, and they constitute microelements. These include iron, manganese, zinc, boron, copper and molybdenum; of these, iron seems to be more critical since chelated forms of iron and zinc are commonly used in preparing culture media. Iron tartrate and citrate are difficult to dissolve and they frequently precipitate in the medium. The problem may be overcome by using EDTA-iron chelate in place of iron citrate, particularly for embryo induction.

Boron plays a major role in the movement of sugar, water and hormones. It is also involved in nitrogenous metabolism, fruiting and cell division. Lack of Boron produces interesting deficiency symptoms – often deterioration of internal tissues as in heart rot of sugar beets (*Beta vulgaris*), cracked stem of celery (*Apium graveolens*), or monkey face in olives (*Olea europaea*). Other symptoms include tip dieback or thick, curled, brittle, chlorotic leaves. Excessive amounts of boron can cause plant injury or death; thus, some herbicides are borates. Boron is added to tissue culture media in small amounts as boric acid.

Chlorine helps in promoting photosynthesis; therefore, it is necessary for the growth. The deficiency syndromes are wilted leaves that become yellowed or bronze and die. Plants require chlorine in only minute quantities, but it is included in some tissue culture media in large amounts as calcium chloride.

Cobalt is essential for nitrogen fixation (the conversion of atmospheric nitrogen to nitrates by means of nitrogen-fixing bacteria). Cobalt chloride is added to most media in amounts of 0.025 mg/l.

Copper is believed to be necessary in energy conversion as it alternates between the cuprous (monovalent) and the cupric (divalent) state. It is involved in chlorophyll synthesis and is found in some enzymes. The deficiency results in stunted growth, malformations, twisted and blotched leaves, or dieback of young twigs. Only 0.025 mg of copper sulphate per litre medium is required to supply the necessary copper in tissue culture of most plants.

Iodine is added to media as potassium iodide and is not considered as an essential element, even though it is a component of some amino acids. Iodine appears to have an adverse effect on *Rhododendron* cultures.

Manganese is required for the enzyme reactions, particularly, in respiratory and photosynthetic processes. The deficiency of manganese results in yellowing of the leaves and is an essential element in chloroplast membrane. Manganese sulphate supplies the necessary manganese for the tissue culture media.

Molybdenum helps to convert nitrogen to ammonia and aids in nitrogen fixation. It is required for normal growth and protein synthesis. Lack of molybdenum will cause whiptail condition in cauliflower or it can stunt the growth of legumes. It is added to the tissue culture media as sodium molybdate, also known as molybdic acid sodium salt. Quantities exceeding 10 ppm can be injurious to the plants.

Zinc is an enzyme activator and is involved in chlorophyll formation. It also helps in the production of the auxin IAA. In the absence of zinc, roots may be abnormal and leaves may turn mottled bronze or yellow. A trace of zinc is included in the tissue culture media as zinc sulphate. Large quantities of zinc are toxic to the plants.

Iron is involved in chlorophyll synthesis and it also participates in energy conversion during photosynthesis and respiration. It is reduced from the ferric (trivalent) to ferrous (divalent) state. Plants deficient in iron have yellowed (chlorotic) young leaves, especially between the veins, giving striped appearance. In tissue culture media, ferrous sulphate is often mixed with the sodium salt of EDTA to sequester the iron, thereby, making it more readily available to the plants. Iron is sometimes classified as a minor element.

2.2.2. Carbon and Energy Source

Sugar is the most important part of a nutrient medium as an energy source, since most cultures are unable to photosynthesize effectively owing to inadequate cellular and tissue development, lack of chlorophyll, limited gas exchange and carbon dioxide in tissue culture vessels etc. Hence, tissue cultures lack autotrophic ability and they need external carbon for their energy. The most preferred carbon source in plant tissue culture is sucrose at a concentration of 20–60 g/l. Sucrose is cheap, easily available, readily assimilated and relatively stable, and is the most commonly used carbon source. Glucose supports equally good growth while fructose is less efficient. Sucrose or glucose at 2–4% is the most suitable carbon source, which is added to the BM. Other sugars such as fructose (toxic if autoclaved), maltose, galactose, raffinose and sorbitol also support the growth of various plant cells, and these may sometimes prove superior to sucrose. These sugars are used depending upon the plant species. During autoclaving of the medium, sucrose is hydrolyzed to glucose and fructose, which are then used up for growth. The most suitable carbon source and its optimal concentration should be chosen to establish the efficient production of useful metabolites, and these factors depend on plant species and products.

- **Vitamins:** Vitamins are organic substances required for metabolic processes as cofactors or parts of the enzymes and as such for optimum growth. Thus, the medium should be supplemented with vitamins. Plants synthesize vitamins endogenously and they are used as catalysts in various metabolic processes. When plant cell and tissues are grown *in vitro*, some essential vitamins are synthesized, but only in suboptimal quantities. Hence, it is necessary to supplement the medium with required vitamins and amino acids to achieve best growth of the tissue. Thiamine (B_1), nicotinic acid (B_3), pyridoxine (B_6), calcium pantothenate (B_5), and myoinositol are used more often; of these vitamins, Thiamine is the basic vitamin required by all the cells and tissues. Nicotinic acid and pyridoxine are usually added to the culture media, but may not be essential for cell growth in many species.

- **Amino acids:** Cultured tissues are normally capable of synthesizing the amino acids necessary for various metabolic processes. In spite of this, the addition of amino acids to media is important for stimulating cell growth in protoplast cultures and establishing cell lines. Unlike inorganic nitrogen, amino acids are taken up more readily by the plant cells. The CH, L-glutamine, L-aspargine, L-glycine, L-arginine and L-cysteine are common sources of organic nitrogen used in the culture media. Tyrosine should be used only when agar is added to the medium. Only L-isomers of the amino acids should be used.

- **Complex organic supplements:** In order to stimulate the growth of the cells, culture media are often supplemented with a variety of organic extracts. These include protein (casein) hydrolysate, CM, yeast and malt extracts, ground banana, orange juice and tomato juice. Similarly, potato extract has also been found as a suitable additive for anther culture. Yeast extract is a good source of organic nitrogen and vitamins.

The CH (Edamin) is an undefined protein mixture that is occasionally used in media as a non-specific source of organic nitrogen. Casein, a phosphorous protein in milk, is hydrolysed (treated with water) to form this weak acid.

The CM is the liquid endosperm of coconut. This undefined medium was used successfully in early plant tissue culture of *Datura* embryos. Caplin and Steward (1948) grew carrot phloem explants using CM and CH as basal salt supplements. Steward achieved eighty-fold multiplication of carrot roots in 3 wk using CM. The CM has also been used widely in orchid culture. In the early 1960s, Carl Withner first cultured *Phalaenopsis* orchids on Vacin and Went's medium, which contains 15% CM.

The CM is available as coconut water from many commercial suppliers, or it can be extracted from fresh coconuts. To do this, select green coconuts that have reached full size (as young or as fresh as can be found in the store). Shake the coconut before you buy it to make sure there is liquid in it. Collect the liquid after punching two holes in the eyes with a clean nail. Use several coconuts at a time to avoid variations between small batches. Deproteinize the water by heating (80°C), filter through Whatman No. 1 paper and store in deep freeze. It is used in the concentration of 5–20% (v/v), which is found to stimulate shoot proliferation in many species of plants.

3.2.3. Antioxidants

The antioxidants such as citric acid, ascorbic acid, pyrogallol, phloroglucinol and L-cysteine are used in tissue culture to reduce excessive browning of the explants. Adsorbents such as PVP and AC are also used for checking excessive browning.

Activated charcoal

The addition of AC to the culture media is reported to stimulate growth and differentiation in orchids, carrot, ivy and tomato. However, it is inhibitory in tobacco, soybean and *Camellia*. Inhibition of the growth is

Preparation of 1 Liter Murashige and Skoog Medium

Take 500 ml of double distilled water in 1 Liter Erlenmeyer Flask

↓

Weigh, add and dissolve the following macro salts one by one in sequence

$MgSO_4.7H_20$	370 mg
KH_2SO_4	170 mg
NH_4NO_3	1650 mg
KNO_3	1900 mg
$CaCl_2.2H_2O$	440 mg
Myo inositol	100 mg

Weigh, add and dissolve 3% Sucrose

↓

Micro Stock-I (40X)

	mg/100 ml
H_3BO_3	248
KI	33.2
$Na_2MoO_4.H_2O$	10
$CoCl_2.6H_2O$	1

Micro Stock-II (40X)

	mg/100 ml
$MnSO_4H_2O$	892
$ZnSO_4.H_2O$	344
$CuSO_4.5H_2O$	1

Micro Stock-III (40X)

	mg/100 ml
$FeSO_4H_2O$	1114
Na_2EDTA	1494

Micro Stock-IV (40X)

	mg/100 ml
Nicotinic acid	20
Pyridoxine	20
Thiamine HCI	20
Glycine	80

Add 2.5ml of each micro stock I, II, III & IV

↓

PGR Stock (0.01%)

10 mg of Cytokinins or auxins or gibberellins is dissolved in suitable solvent and made up to 100 ml

Add aliquots of PGR stock of desired concentration

↓

Add DDW up to 900ml in Erlenmeyer flask

↓

Adjust pH to 5.8

↓

Pour medium into graduated cylinder and make up to 1 liter

↓

Add 0.8% agar agar

↓

Melt agar agar with intermittent shaking

↓

Dispense the medium of 15ml each into culture tubes or 50 ml each into culture bottles

↓

Autoclave at 15 psi at 121°C for 15 min

Fig. 7 Preparation of MS medium

attributed to the adsorption of phytohormones by AC, whereas stimulation could be due to adsorption of inhibitory compounds by AC. AC is generally acid-washed and neutralized, before its addition to the culture medium, at the concentration of 0.5–3%. Being porous, AC adsorbs tissue exudates in the culture and helps to reduce the toxicity by removing toxic compounds (e.g., phenols) produced during the culture.

3.2.4. Plant Growth Regulators

The PGRs are the critical media components in determining the developmental pathway of the plant cells. These stimulate cell division, and regulate the growth and differentiation of the shoot and the root on explants and embryos in semi-solid and liquid medium cultures. The most commonly-used growth regulators are phytohormones and their synthetic analogues. Phytohormones are required to induce the callus tissue and also to promote the growth of many cell lines. Since each plant species requires different kinds and levels of phytohormones for the induction of the multiple shoots, callus and its growth and metabolite production, it is important to select the most appropriate growth regulators and to determine their optimal concentrations. The concentration and ratio of the hormones may vary from plant to plant, and should be standardized for a particular plant tissue. Four broad classes of growth regulators, *viz*., auxins, cytokinins, gibberellins and ABA or their synthetic counterparts (either singly or in combination) are important in tissue culture. The growth regulators used in the cultures are categorized into 5 main classes, *viz*., auxins, cytokinins, gibberellins, ABA and ethylene.

Auxins

- IAA
- NAA
- IBA
- 2,4-D
- NOA

Others

- 4-Chlorophenoxy acetic acid
- 2,4,5 -T

- MCPA
- Picloram

The 2,4-D and 2,4,5-T are commonly used for callus production.

Auxin

An auxin (IAA, NAA, 2,4-D, 2,4,5-T or IBA) is required for the induction of cell division and root initiation in cultured tissues. Auxin causes cell division, cell elongation and swelling of the tissues, and the formation of adventitious roots. At low auxin concentrations, adventitious root formation predominates, whereas at high auxin concentrations, root formation fails to occur and callus formation takes place. The auxins are mostly used in combination with cytokinins. 2,4-D and 2,4,5-T are mostly used for callus induction whereas IAA, IBA and NAA are used for root induction. In nature, the hormones of this group are involved with such activities as elongation of stem, internodes, tropism, apical dominance, abscission and rooting. Auxins are thermostable; however, IAA is destroyed by low pH, light, oxygen and peroxidase. Various auxins differ in their physiological activity and the extent to which they move through the tissue or are bound to the cells. Based on the stem curvature assays, 2,4-D has 8–12 times the activity of IAA compared to 2,4,5-T, PCPA and Picloram, which are four times or twice as effective as IAA. The NAA and 2,4-D are most stable forms of auxin.

Cytokinins

- BAP (synthetic)
- BA (synthetic)
- 2iP (natural)
- KN (synthetic)
- Zeatin (natural)

The cytokinins (KN, BAP, 2iP and Zeatin) are adenine derivatives and they promote cell division, shoot proliferation, organogenesis and somatic embryogenesis. They have essential role in differentiation and micropropagation of most plant species. The most frequently used cytokinins are KN, BA or BAP, Zeatin, and 2iP. Zeatin and 2iP are naturally occurring cytokinins while BAP and KN are synthetically derived cytokinins. These are often used to stimulate growth and development. They usually promote cell division if added together with an auxin. At higher concentrations (1–10 mg/l), adventitious shoot formation is induced, but root formation is generally inhibited. They promote axillary shoot formation by decreasing apical dominance.

It was subsequently discovered that DPU, thidiazuron, CPPU and other derivatives of DPU show cytokine-type activity (Pierik, 1989).

The ratio of auxins and cytokinins is important with respect to morphogenesis in the culture system. For embryogenesis, callus initiation and root initiation the requisite ratio of auxins to cytokinin is high while the reverse leads to axillary and shoot proliferation.

Gibberellins

These are infrequently used in plant tissue cultures as it can inhibit the callus growth. Gibberellins are normally used in plant regeneration. GA_3 is essential for meristem culture of some species. In general, gibberellins induce elongation of internodes and the growth of meristems or buds *in vitro*. Gibberellins usually inhibit adventitious root as well as shoot formation.

The ABA

This is an important growth regulator for induction of embryogenesis. It is also known to promote morphogenesis in *Begonia* cultures. Ethylene is produced by the cultured cells, but its role in cell and organ culture is not known. It may be promotory or inhibitory for the same process in different systems. For example, it promoted somatic embryogenesis in maize, but the same process was inhibited in *Hevea brasiliensis*.

Ethylene

This is a gaseous, naturally-occurring growth regulator associated with controlling the ripening of fruit in climacteric fruits. Its use in cultures is not widespread. Some cultures generate ethylene, which if it builds up sufficiently, can inhibit the growth and development of the culture.

3.2.5. Other Growth Regulators Useful in Plant Tissue Culture

Polyamines

Polyamines are compounds that occur in high levels within the plants and are used in tissue culture media at concentrations of 10–1000 mM.

Polyamines may enhance regeneration of the roots, shoots and embryos; delay or prevent senescence and regulate flowering. The enhancement of morphogenesis by polyamines may be through their abatement of ethylene production. Putrescine and spermidine are the polyamines.

Jasmonic acid

The JA can influence several aspects of the plant growth and development. The JA and its methyl ester, JAMe, are naturally-occurring regulators of higher plant development, responses to external stimuli, and gene expression. Jasmonate can induce senescence in leaf. The JA was first isolated from cultures of the fungus *Lasiodiplodia theobromae* (Demole *et al.,* 1962). Its volatile methyl ester is involved in wound and pathogen responses, formation of stress proteins, and ethylene synthesis (Gaspar *et al.,* 1996). It stimulates bulb formation *in vitro* (Gaspar *et al.,* 1996), retards callus formation and promotes bud formation in potato meristems (Ravnikar and Gogala, 1990). It also inhibits the shoot formation causing fewer and shorter shoots from *Pinus radiata* cotyledons (Gaspar *et al.,* 1996). It plays an imporant role in the enhancement of secondary plant metabolites, and observed increased taxane accumulation in callus culture of *Taxus x media* (Furmanowa *et al.,* 1997)

There are several other growth regulators such as brassinosteroids; steroid class compounds that promote cell expansion and elongation, and salicylic acid, a phenolic phytohormone that enhances regeneration in *Coffea arabica* (Quiroz-Figueroa and Mendez-Zeel, 2001).

3.2.6. Gelling Agents

If liquid medium is used in static cultures, the tissue would get submerged and die due to lack of oxygen. A gelling agent is generally used to circumvent this problem. The most desirable property of a gelling agent is that it should withstand sterilization by autoclaving, and the medium should be liquid when hot, but form a semi-solid gel when cool. Plant tissues have been shown to do better when cultured on a semi-solid support than otherwise. Most of the gelling agents (agar, agarose, gelrite) used in the plant tissue culture media are biological products. Agar is the most commonly used gelling agent and is obtained from red algae, especially *Gelidium amansii.* It is a complex mixture of related polysaccharides built up from the sugar, galactose. These include the neutral polymer fraction, agarose, which gives strength to the gel and the highly-charged anionic polysaccharides, agaropectins, which give viscosity to agar.

As a gelling agent, the agar in tissue culture medium is strong enough to support the culture; yet liquid enough to allow the nutrients to diffuse through the medium to the plantlets. Agar up to 6–8 g/l of the medium is used.

When greater purity is needed, agarose may be used. Agarose is extracted from agar leaving behind agaropectin and its sulfate groups. Due to the additional purification, agarose is considerably more expensive than agar. Agarose also has higher gel strength than agar and thus, less is required for the solidification of media. Agarose is used in situations where the impurity of agar is a major disadvantage, such as in protoplast culture.

Gelrite consists of a polysaccharide produced by bacterium *Pseudomonas elodea*. Medium solidified with Gelrite has an advantage of being clear, which the agar-solidified medium does not. Consequently, contamination is more easily detected at an early stage. Impurities in Gelrite contain inorganic ions, but no organic compounds. Gelrite requires more stirring than agar when being added to the media. Unlike agar, Gelrite cannot be reheated and gelled successfully. One limitation of Gelrite is that the concentration of divalent cations such as calcium and magnesium ions must be within the range of 4–8 mM/l. Concentrations of these two ions either less than or greater than this range result in the media not gelling. When Gelrite is used at low concentrations, it may also produce hyperhydric plants.

Phytagel is an agar substitute produced from a bacterial substrate composed of glucuronic acid, rhamnose and glucose. It produces a clear, colourless, high-strength gel, which aids in the detection of microbial contamination. Phytagel provides an economical alternative to agar as a gelling agent. It is used at a concentration of 1.5–2.5 g/l. Phytagel should be added rapidly to stirring-culture medium at room temperature to prevent clumping. Hyperhydricity may also be a problem with this gelling agent.

Mechanical supports such as filter paper folded into wicks and polyethylene rafts can be used with liquid medium to ensure adequate supply of oxygen. Many other materials have been used as well including rock wool, cheese cloth, sand, and pieces of foam. Explants can be grown in rotating cultures where tissue is alternately bathed in liquid and exposed to air.

3.3 MEDIA PREPARATION

Stock solutions (40X) of micronutrients, vitamins and growth hormones are prepared separately, and stored in a refrigerator and can be used up to 2 mo. Iron stock is prepared separately as chelated form as the sodium salt of the ferric EDTA and stored in amber-coloured bottle in a refrigerator. Different solvents are used for the preparation of stock solutions of hormones; concentrated Hydrochloric acid (HCl) for cytokinins (BAP and KN), 1 N sodium hydroxide (NaOH) for auxins (IAA, IBA and NAA), ethyl alcohol (EtOH) for 2,4-D, 2,4,5-T and 2,4,5-TP and DMSO for TDZ (Table 4). Auxin stocks are usually prepared by dissolving 1 N sodium hydroxide or 1 N potassium hydroxide (KOH) in ethanol, until crystals are dissolved (not more than 0.3 ml/10 mg of auxin), rapidly adding 90 ml of distilled water and increasing the volume to 100 ml in a volumetric flask. Cytokinin stocks are prepared in a few drops of 1 N hydrochloric acid and water to dissolve crystals. Gentle heating is usually required to complete dissolving of crystals. The stock is made up to the desired volume by adding double distilled water in a volumetric flask. Cytokinin stocks can be stored for several months in the refrigerator. Cytokinins are thermostable during autoclaving in media.

Table 4: Solvents used for dissolving different PGRs

Growth regulator category	Name of the PGR	Solvent
Auxin	2,4-D	1 N Sodium hydroxide
	IBA	1 N Sodium hydroxide
	NAA	1 N Sodium hydroxide
	Dicamba	1 N Sodium hydroxide
	Picloram	1 N Sodium hydroxide
	IAA	Ethyl alcohol /1 N Sodium hydroxide
	2,4,5-T	1 N Sodium hydroxide
	2,4,5-TP	1 N Sodium hydroxide
Cytokinin	BAP	1 N Sodium hydroxide
	KN	1 N Sodium hydroxide
	TDZ	DMSO or potassium hydroxide
Gibberellin	GA_3	1 N Sodium hydroxide

The macronutrients are prepared in concentrations of 5X, 10X etc. depending upon the frequency of usage. While preparing macronutrient stocks, in order to avoid the precipitation, calcium (supplied in the form of nitrate or chloride) is prepared separately.

After weighing all the constituents of the medium properly, they are dissolved in double-distilled water. Appropriate quantities of the stock solutions are mixed together to prepare required volume of the BM. Hormonal stocks are added in required concentrations either individually or in various combinations. After the thorough mixing of the contents in water, sugar and other organic supplements are added. Finally, the volume is made up to 1 l. The pH of the medium is adjusted around 5.8 (slightly acidic) by adding 0.1 N sodium hydroxide or 0.1 N hydrochloric acid. In order to make the medium semi-solid, appropriate quantities of agar is added to the liquid medium and boiled till the agar melts totally. Now, equal quantities of the medium are dispensed into the culture vials (test tubes, flasks or petri plates). The culture vials containing the medium are plugged either with non-absorbent cotton wrapped in cheesecloth, aluminium foils or plastic closures, and autoclaved under 15 psi at 121°C for 15–40 min. Culture medium is allowed to cool and then stored at 4°C until use.

It is desirable to make slants while preparing a solid medium in culture tubes by keeping the tubes in tilted position during cooling. This is because the slants provide larger surface area for tissue growth.

3.3.1 Aeration

The tissue or organs can be cultured either on a solid medium or liquid medium. Agar-agar (a polysaccharide obtained from sea weeds) is used at a concentration of 0.8–1% to solidify the medium. When liquid medium is used, aeration is provided by shaking the vials on an automatic shaker. The advantage of growing the tissues in agitated liquid media is that it facilitates gaseous exchange and reduces nutrient gradient.

3.3.2 Summary of the Medium Preparation

1. Prepare stock solutions of the major inorganic nutrients, trace elements, vitamins and growth regulators.
2. Pipette the required volume of each stock solution into a 1 *l* glass beaker to make up 1 *l* of the liquid medium and keep it on a magnetic stirrer.

3. Add myoinositol and sucrose as solids and allow these to dissolve fully adding more water if necessary.
4. Make the volume approximately up to 950 ml with distilled water.
5. Adjust the pH to 5.8 (slightly acidic) with 0.1 N sodium hydroxide or 0.1 N hydrochloric acid.
6. Transfer the medium to 1 *l* measuring cylinder or volumetric flask and make up to 1 *l* with distilled water.
7. Transfer the medium back to the stirred flask to allow full mixing.
8. Boil the medium to melt the agar.
9. Transfer the medium to culture vials (flasks, test tubes, petri plates).

3.4 COMMONLY USED CONCENTRATIONS IN TISSUE CULTURE

Concentrations of any substance can be given in several ways and are interconvertable.

- **Molar solution:** Molar concentrations are used to compare the relative activity among different compounds accurately. A mole is the same number of grams as the molecular weight (Avogadro's number 6.02×10^{23}) of the molecules. Hence, a 1 molar solution represents 1 mol of the substance in 1 l of the solution and 0.01 M represents 0.01 times the molecular weight in 1 l. A millimolar (mM) solution is 0.01 mol/l and micromolar (µM) solution is 0.000001 mol/l. The PGRs are active at the micromolar concentrations. 1 µM substance means concentration of IAA would contain the same number of molecules as a 1 µM concentration of KN, though the same could not be said for units based upon weight.
- **Milligrams per litre (mg/l):** 1 mg is 10^{-3}g. 1 mg/l means placing 1 mg of the desired substance in a final volume of 1 l of solution.
- **Micrograms per litre (µg/l):** 1 µg = 0.001 or 10^{-3} mg = 0.000001 or 10^{-6} g. It means placing 1 µg of substance in 1 l of solution.

- **Parts per million (ppm):** Media components are expressed in ppm, which means 1 part per million or 1 mg/l.
- **Percentage based upon volume% (v/v):** It is used for additive substances such as CM, sugarcane juice, tomato juice and if a 100 ml of a 10% (v/v) CM was desired then 10 ml of CM would be diluted in 100 ml of water.
- **Percentage based upon weight% (W/V):** This is often used to express concentrations of agar or sugar. Dissolve 10 g of sucrose in 1 l of medium to make 1% (W/V) sucrose solution.

■■■■

4 METHODS OF STERILIZATION

4.1. INTRODUCTION

The application of plant cell, tissue and organ culture requires sterile technique. The maintenance of aseptic or sterile conditions is essential for successful tissue culture procedures. A few simple precautions to avoid contamination will save the valuable time in avoiding the repeated experiments. Sterilization is a procedure used for elimination of the microorganisms.

The most important factor in the selection of suitable working area is the possible flow or unfiltered air over the disinfected working area; air currents must be avoided, because of air-borne spores or contaminating microorganisms.

For aseptic manipulation, the instruments such as forceps, scalpels and scissors are sterilized by dipping in ethanol, followed by flaming and cooling, which is referred to as flame sterilization.

The methods of sterilization can be classified as follows:

Physical methods

- Heat (dry heat, moist heat)
- Microfiltration
- Irradiation

Chemical methods: strong disinfection e.g., formaldehyde, ethylene oxide etc.

4.2. STERILIZATION BY HEAT

Fire and boiling water have been used for sterilization and disinfection since the time of Greeks, and heating is still one of the most popular ways to destroy the microorganisms. Either moist heat or dry heat does this. Moist heat is the most effective and efficient than dry heat. It kills the organisms by coagulating and denaturing their enzymes and proteins; this is helped a lot by the presence of water. Spores are killed by exposure to moist heat at 121°C for 10–30 min. Dry heat kills the organisms by oxidative destruction of the cell constituents. Spores are killed by exposure to dry heat at 160°C for 1 h. Following factors influence sterilization by heat:

- Temperature and time of exposure
- The number of vegetative organisms and spores present
- Nature of the organism

4.2.1 Thermal Death Point

As heat is very useful in controlling the microorganisms, it is essential to have a precise measure of the heat-killing efficiency. The TDP is the lowest temperature to give complete killing in an aqueous suspension within 10 min at standard conditions. As TDP implies that a certain temperature is immediately lethal despite the conditions, TDT is now more commonly used. This is the shortest time needed to kill all the organisms in a microbial suspension at a specific temperature under defined conditions. However, such destruction is logarithmic and it is theoretically not possible to destroy the microorganisms in a sample, even with extended heating. Therefore, an even more precise figure, the DRT or D-value has gained wide acceptance. This is the time in minutes required to achieve a ten-fold reduction in viability of a bacterial population, at a given temperature under standard conditions.

Table 5: Conditions for moist-heat killing

Organism	Vegetative cells	Spores
Yeasts	5 min at 50–60°C	5 min at 70–80°C
Moulds	30 min at 62°C	30 min at 80°C
Bacteria	10 min at 60–70°C	2–over 800 min at 100°C
Virus	30 min at 60°C	0.5–12 min at 121°C

4.2.2. Sterilization by Dry Heat

This method is used only for glassware, metal instruments, and other materials that are not charred by oven temperatures. Strips of tape should not be on any glassware, and prolonged oven heat may dull the cutting edges of surgical blades. Cotton, paper, and plastic should never be placed in an oven for sterilization. It is highly important not to use a laboratory oven that has previously been used for paraffin embedding.

Objects to be sterilized are wrapped in heavy-duty aluminium foil before being placed in the oven. Care should be taken not to pack the foil wrapped packages in the oven too tightly, but to leave some space between them. After sterilization, the wrapped packages are taken to the laminar flow cabinet. Three time periods must be considered in calculating the time required for dry heat sterilization.

1. Approximately 1 h (heating-up period) is allowed for the entire load to reach the sterilization temperature of 180°C.
2. A minimum of 2 h is required to kill all the organisms including the spore formers at this temperature.
3. Finally, a cooling-down period is advisable to prevent the glassware from cracking due to rapid drop in the temperature.

Normally, the instruments such as forceps, scalpels, needles and spatulas are sterilized by dipping in 95% ethyl alcohol, followed by flaming and cooling. This technique is called flame sterilization. The instruments are repeatedly sterilized during the process to avoid contamination. It is customary to flame the mouths of the culture vessels prior to inoculation or subculture.

4.2.3. Sterilization by Moist Heat

This procedure employs an autoclave operated with steam under pressure. The culture vials containing medium are plugged either with non-absorbant cotton wrapped in cheesecloth, aluminium foils or plastic closures and autoclaved under 15 psi (1.06 kg/cm^2) and 121°C for 15–40 min. The time required for the sterilization of the liquids varies considerably depending on the volume. The greater the volume, the longer it will take for the content to reach the sterilization temperature.

4.3. FILTER STERILIZATION

Filter sterilization is the process of removing contaminants in the range of 0.025–10 μm from fluids by passage through microporous medium such as membrane filter. This is necessary for media components that are degraded during autoclaving and must be sterilized at room temperature. A relatively small volume can be sterilized by passage through filtration unit attached to a hypodermic syringe. Most of the membrane filters are screens consisting of a uniform continuous mesh of polymeric material with pore size precisely determined by the manufacturing process. Although some workers use a 0.45 μm pore size, the pore diameter should measure at least 0.22 μm for the complete removal of all bacteria, yeast, and moulds.

4.4. CHEMICALS

The working area is generally disinfected with either ethanol or isopropanol (70% v/v). Although acidified alcohol (70% v/v, pH 2.0) may be a more effective disinfectant, it has a corrosive effect on metal instruments.

- **Sodium Hypochlorite (NaOCl):** It is the most-common chemical agent (0.02–0.25%) used to sterilize the plant tissues. Diluted household commercial bleach can also be used for this purpose, which contains 5.25% sodium hypochlorite. It is less expensive and equally effective.

- **Calcium Hypochlorite (CaOCl):** It can be used as a substitute for sodium hypochlorite. However, calcium hypochlorite causes slightly less damage to the plant tissues. In order to avoid accumulation of calcium hypochlorite on the plant tissue surfaces, sterilization solutions should be filtered or decanted prior to use.

- **Hydrogen Peroxide (H_2O_2):** Plant tissues can also be surface-sterilized using 3–10% hydrogen peroxide. It is much easier to sterilize the tissues with hydrogen peroxide than sodium hypochlorite and calcium hypochlorite.

- **Other Substances:** Bromine water (1–2%), silver nitrate ($AgNO_3$) 1% and mercuric chloride ($HgCl_2$) 0.1–1% can also be used for surface sterilization of the tissues.

4.5. INFRARED RADIATION

The infrared rays are directed from an electricity-heated element on to the objects such as glass syringes, etc. at 180°C. Surgical instruments are sterilized above 200°C in vacuum chamber in which molecular nitrogen is passed during cooling time to prevent oxidation.

4.5.1. Sterilization by UV Radiation

The UV radiation around 260 nm is quite lethal, but does not penetrate glass, dirt films, water and other substances very effectively. Due to this disadvantage, UV radiation is used as a sterilizing agent only in a few specific situations. The UV lamps are sometimes placed on the ceilings of rooms in biological safety cabinets to sterilize the air and any exposed surfaces.

The UV radiations are harmful to mankind; therefore, skillful installation is required during the practical use. The UV radiation burns the skin and damages the eyes. People working in such areas must switch off the UV lamps when the areas are in use. Commercial UV units are available for water treatment. Pathogens and other microorganisms are destroyed when a thin layer of water is passed under the lamps. The UV rays kill many of the vegetative forms of microorganisms present in the air, but they vary in their susceptibility. These radiations are being used in sterilizing the surgical area, instruments, hospital, school, storage warehouses etc. The UV rays act by inducing the dimers in DNA, and this interferes with the replication of DNA of the organisms. Mercury vapour lamps produce these UV rays. It is used to sterilize the plastic material.

4.6. SURFACTANTS

- **Tween 20 or Triton X-100:** These are the scientific reagent-grade surfactants often used in low concentrations (0.05%) to ensure that the sterilizing agent come in contact with the entire plant tissue surface.
- **Stirring of the tissues:** Good surface contact is facilitated by stirring the tissues during surface sterilization.
- **Ultrasonic bath:** Ultrasonic bath is used to clean the dentures, and it is an effective method to ensure good surface contact during sterilization treatment.

These are particularly useful for sterilizing buds and woody tissues that have many small-surface crevices and cracks. However, minimum of three sequential rinses with sterile distilled water are recommended to remove the traces of these chemical sterilants.

4.7 ANTIBIOTICS FOR CONTROL OF BACTERIAL AND FUNGAL CONTAMINANTS

Antibiotics are used to reduce, control and eliminate the contaminants in plant tissue cultures. Most antibiotics are heat sensitive and cannot be autoclaved. They are dissolved in water or another suitable solvent, filter-sterilized and added to autoclaved culture media that has cooled to 45–50°C. Falkiner (1990) suggested the list of antibiotics, which are suitable for the plant tissue culture.

Table 6: Antibiotics and their mode of action

Antibiotics	Mode of action
Aminoglycosides Kanamycin Neomycin Streptomycin	Inhibits protein synthesis by interaction with 30S and 50S ribosomes
β-Lactams Ampicillin Carbenicillin Penicillin	Inhibits bacterial cell wall synthesis
Chloramphenicol	Inhibits protein synthesis by acting on 50S ribosome
Glycopeptides Vancomycin	Interferes with bacterial cell wall synthesis
Macrolides and lincosamides Erythromycin Lincomycin	Inhibits protein synthesis by acting on 50S ribosomes
Rifampicin	Interferes with mRNA formation by binding to RNA polymerase
Tetracyclines	Inhibits protein synthesis by acting on 30S ribosome
Trimethoprim and Sulphonamides	Inhibits synthesis of tetrahydrofolate

4.8. MAINTENANCE OF ASEPTIC CONDITIONS

Prior to commencing work, the hands are washed with aseptic soap followed by swabbing with ethanol. The laminar airflow chamber is sprayed with 95% ethyl alcohol. All the paraphernalia such as medium, forceps, scalpel, petri plates etc. used for inoculation are transferred to the laminar airflow chamber and UV light is switched on for 30 min. Since UV radiation is harmful to the eyes, care should be taken to avoid direct exposure to UV. After half an hour, UV light is switched off and air flow is switched on, which flows at the velocity of 27 ± 3 m/min.

Work in the laminar is started after 15 min of air flow to remove the ozone (O_3), which was built up in the chamber during UV radiation. In order to minimize the contamination during inoculation, hands and inoculating area are swabbed with alcohol frequently, followed by flame sterilization of the forceps and scalpel.

Tie back your hair, roll your sleeves up, and remove your watch and other jewellery. Wash your hands thoroughly with the disinfectant solution suitable for skin application, and start the inoculation.

4.9. SURFACE STERILIZATION OF EXPLANT

The explant material, which is used for inoculation, is generally loaded with many microorganisms. The growth of explant will be hampered because the microorganisms with brief life cycle grow very fast on the medium. Hence, the plant material is cleaned and surface sterilized for the development of explant without any problem.

The material is first washed with a detergent followed by washing with double-distilled water several times prior to sterilization. Buds and twigs are cleaned by immersing them briefly in a 70% alcohol solution, which wets and spreads over the tissue surfaces more effectively. A sterilant such as sodium hypochlorite (0.5–5%), or hydrogen peroxide or mercuric chloride (0.1–1.0%) solution may be used for removal of the microorganisms and fungal spores (Table 7). The so treated material is rinsed with sterilized double-distilled water several times to completely wash off the sterilant.

The concentration of the sterilant and the duration of exposures are dependent on the explants. When the seeds are taken as explants, higher concentration can be used.

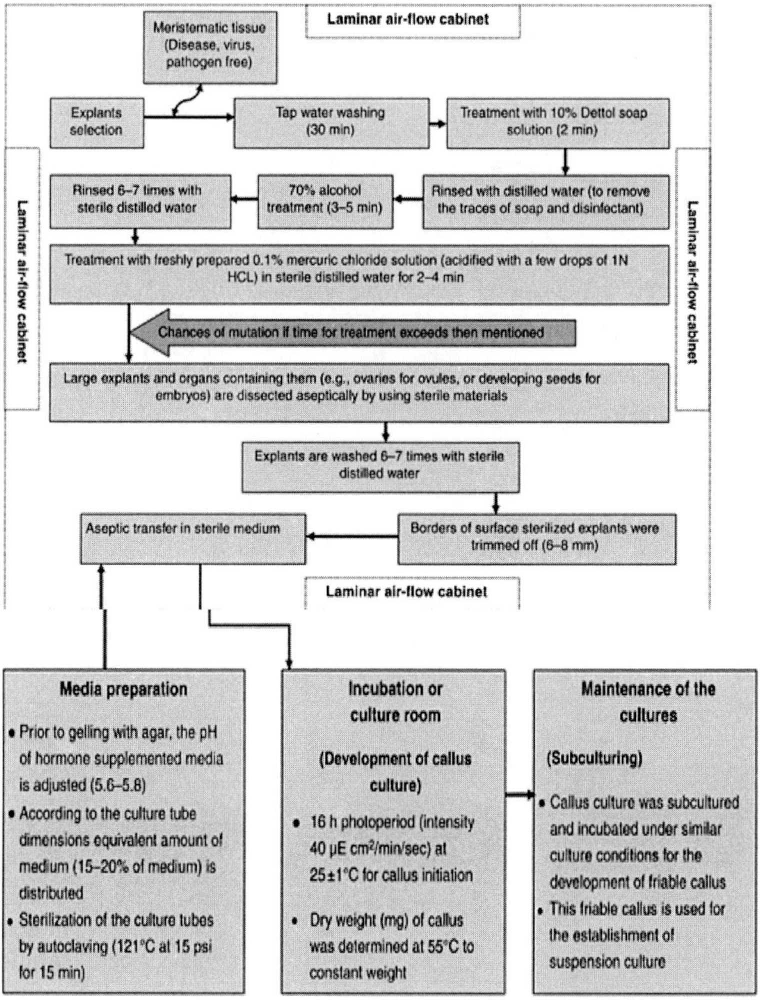

Fig. 8 Steps involved in surface sterilization of the explants

4.10. TRANSFER OF AN EXPLANT

Inoculation is carried out near the spirit lamp. Sterilized plant material is transferred on to a sterilized petri plate containing filter paper. With the help of forceps and scalpel, the explants are dissected and the damaged ends are removed, and blotted on a sterile filter paper. Then the explants are inoculated on to a culture medium vertically to maintain the polarity.

The mouth of the culture tubes are covered with sterilized aluminum foil and tied with rubber bands or plugged with non-absorbant cotton. Culture tubes are labelled with the details of the experiment, name of the explant, date of inoculation etc.

Table 7: Treatment with sterilizing agents commonly used in plant cell and tissue culture

Disinfectant	Concentration (%)	Duration of treatment (min)
1. Sodium hypochlorite	0.5–5	5–30
2. Calcium hypochlorite	9–10	5–30
3. Hydrogen peroxide	3–12	5–15
4. Mercuric chloride	0.1–1	2–10
5. Ethyl alcohol	75–95	—
6. Benzalkonium chloride	0.01–0.1	5–20
7. Silver nitrate	1	5–30
8. Bromine water	1–2	2–10

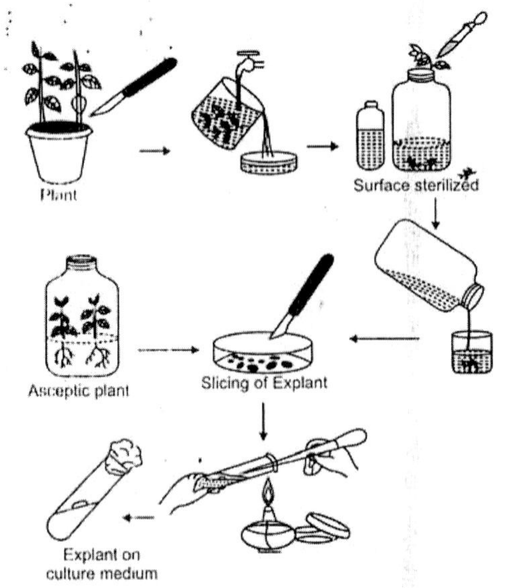

Fig. 9 Diagram showing the procedure for surface sterilization of the plant material and inoculation of the explants for culture

The culture tubes are shifted to the culture room, which has the physical conditions such as light (intensity and duration), temperature and relative humidity maintained at optimum levels.

■■■■

5 CELL AND SUSPENSION CULTURE

5.1. INTRODUCTION

Many higher plants are the major sources of natural products used as pharmaceuticals, agrochemicals, flavour and fragrance ingredients, food additives, and pesticides (Balandrin and Klocke, 1988; Goossens *et al.*, 2003). The search for new plant-derived chemicals through callus induction should, thus, be a priority in current and future efforts towards sustainable conservation and rational utilization of biodiversity.

5.2. PROCEDURE FOR CELL AND SUSPENSION CULTURE

1. Cut internodal or leaf segments into segments (0.5 cm) and culture them on 8% (w/v) agar solidified MS media, supplemented with growth regulators (1-NAA and BAP) at different concentrations and combinations for callus induction.
2. After 12 d, culture the isolated cells of the explants on petri dishes with liquid MS media, supplemented with different concentrations of BAP and combination with NAA. Place the isolated cells on the growth chamber under light for callus induction.
3. After 7 d of culture, the callus is initiated from isolated single cells.

5.3. ESTABLISHMENT OF CELL SUSPENSION CULTURE

The technique of suspension culture is as follows (Fig. 10):

1. Mechanically isolate the cells from the pre-established friable callus tissue by a needle, and subsequent cell suspension culture by a pipette.

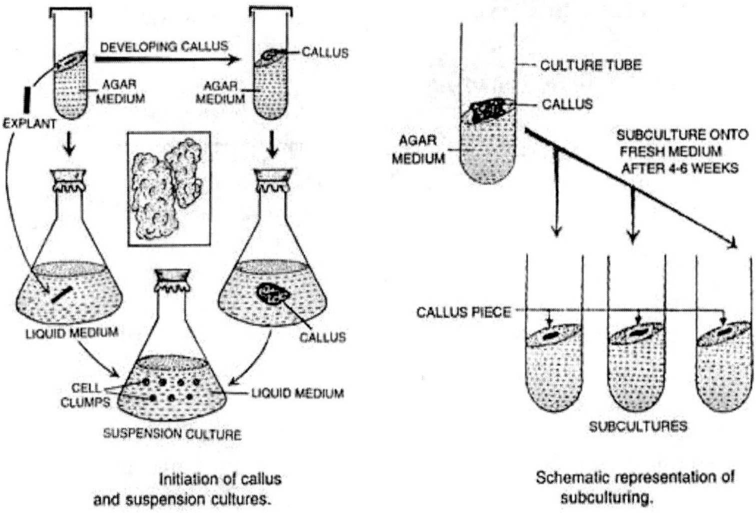

Fig. 10 Technique of suspension culture and subculturing

2. Transfer the friable callus into a liquid medium to make the cell suspension. Continuously agitate the liquid medium by an orbital shaker. Due to continuous agitation, the cell clumps are broken and dispensed releasing single cells in the medium, resulting in cell suspension.

3. After filtration, centrifuge the cell suspension to collect the single cells from the suspension. This procedure is stated below:

   a. Take conical flasks (250 ml) and keep 75 ml of the liquid MS medium each flask, and autoclave at 121°C temperature under 15 psi for 20 min.

   b. After autoclaving, take the conical flasks containing MS liquid media in running laminar airflow. Aseptically transfer 4–5 pieces (about 1.5 g) of pre-established friable greenish calli to the conical flasks having 75 ml autoclaved liquid medium (without agar).

   c. Cover the flasks with aseptical aluminum foil and place them within the clamps of an orbital shaker moving at 200 rpm.

d. After 7 d, pour the contents of each flask through a sterilized sieve, having 500 μm diameter pores, and collect the filtrate in a big sterilized container.

5.4 CELL SUSPENSION CULTURE AND CELL GROWTH

Free cells isolated from calli were grown as single cells under *in vitro* conditions, using MS media supplemented with different concentrations and combinations of growth regulators, to observe the growth pattern of cells. Using the sterilized pipette, 5 ml residue cells are resuspended into the fresh liquid medium and the cells are dispensed equally in several sterilized flasks of 250 ml containing 70 ml liquid medium. The cell-containing flasks are placed on shaker with 250 rpm, allowing the free cells to grow. The weight of the cells are taken at different times, from each flask individually, and fresh weights of the growing cells are weighed according to the filter paper method (Dicosmo and Misawa, 1995; Hossain *et al.,* 2007) to prepare a growth curve of the dividing cells.

5.5. FORMATION OF SOMATIC EMBRYOS AND PLANT REGENERATION

Calli induced from single cells in liquid culture are rescued aseptically and cultured on MS semi-solid medium under light to initiate embryo formation. Callus are subcultured individually in MS semi-solid media (with agar) having only BAP and/or BAP with NAA for the formation of somatic embryo. Number of embryos per callus is counted under stereomicroscope, from the beginning of embryogenesis, and average number of the embryo is calculated. Embryos are transferred to MS semi-solid medium containing different concentrations and combinations of BAP, NAA, IBA and IAA for shoot and root induction.

5.6. TYPES OF SUSPENSION CULTURES

Broadly speaking, there are two types of suspension cultures (Fig. 11):

1. Batch cultures

Batch culture is a type of suspension culture where the cell material grows in a finite volume of agitated liquid medium. Generally, cell suspensions are grown in the flasks (100–250 ml) containing 20–75 ml of the liquid culture medium. Batch suspension cultures are most commonly maintained in conical flasks incubated on orbital platform shakers at the speed of 80–120 rpm.

a. **Shake cultures:** Single cells and cell aggregates in fixed volume of liquid medium are placed in conical flasks. Conical flasks are mounted with the help of clips on a horizontal large square plate of an orbital shaker. The square plate moves by a circular motion at the speed of 60–180 rpm.

b. **Stirred cultures:** This system is used for large-scale batch culture (11.5–10 l). In this method, the large culture vessel is not rotated, but the cell suspension inside the vessel is kept continuously dispersed by bubbling sterile air through the culture medium. The use of internal magnetic stirrer is the most convenient way to agitate the culture medium safely. The magnetic stirrer revolves at 200–600 rpm. The culture vessel is a 5–10 l round-bottom flask.

2. Continuous cultures

The large-scale cultures grown under steady state for long periods by adding fresh medium and draining out the used medium in a number of specially-designed culture vessels are known as continuous or mass cultures. Continuous cultures are of the closed or open type. In the closed type, the addition of fresh medium is balanced by the outflow of old medium. The cells passing through the outflow of old medium are separated mechanically and reintroduced in the culture. Cell biomass continues to increase as the growth proceeds. Paradoxically, the inflow of medium in the open type is accompanied by a balancing harvest of an equal volume of culture medium and cells. This allows the indefinite maintenance of the cultures at a constant and sub-maximal growth rate.

Basically, there are two major types of open (continuous) cultures *viz.*, chemostat and turbidostat.

a. **Chemostat:** In this system, culture vessels are generally cylindrical or circular in shape. These vessels possess inlet and outlet pores for aeration, and for the introduction and removal of the cells and the medium. The liquid medium containing the cells is stirred by a magnetic stirrer. The introduction of fresh sterile medium, which is pumped in at a constant rate into the vessel, is balanced by the displacement of an equal volume of spent or old medium and cells. Such a system can be maintained in a steady state; hence, new cells are produced by division at a rate, which compensates the number lost in the outflow of spent medium. Thus, in a steady state condition, the density, growth rate, chemical composition and metabolic activity of the cells remain constant. Such continuous cultures are ideal for studying growth kinetics and the regulation of metabolic activity in higher plants.

b. **Turbidostat:** The turbidity of a suspension culture medium changes rapidly when the cells increase in number due to their steady-state growth. The changes in turbidity of the culture medium can be measured by the changes in OD of the medium. The pH of the medium changes due to an increase in the cell density. In turbidostats, an automatic monitoring unit is connected with the culture vessel, and such unit adjusts the medium flow in such a way as to maintain the OD or pH at a chosen level.

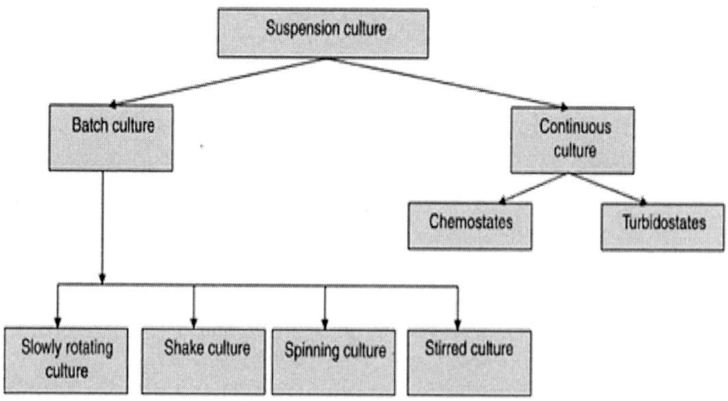

Fig. 11. Schematic representation of the types of suspension cultures

5.7. GROWTH PATTERNS IN SUSPENSION CULTURE

The growth of the suspension culture follows a predictable pattern or growth curve under appropriate light, temperature, aeration and nutrient medium. The growth of suspension culture can be monitored very easily by simply counting the cell number per unit volume of the culture in relation to the days of culture. A typical growth curve can be prepared from such data. The growth curve for a suspension culture consists of lag phase, logarithmic phase or exponential phase, linear phase and stationary phase (Fig. 12).

1. The lag phase is the period where the cells adjust themselves to the nutrient medium, and undertake all the necessary synthesis prior to cell division.
2. This is followed by a very rapid cell division causing a logarithmic increase in cell number. This phase is called logarithmic phase.
3. A further period of rapid cell division results in a linear increase in the number, and the phase is called linear phase.
4. As nutrients are depleted and some of the cells of the culture show senescence characteristics, the rate of cell division within the cell culture declines, and it passes through stationary phase. At this stage, the growth curve forms a plateau.

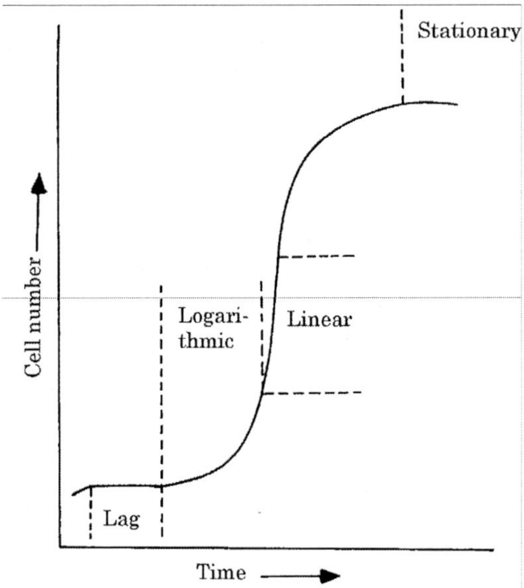

Fig. 12 The growth curve for a typical higher plant suspension culture showing the major phases of growth

5.8. MEASUREMENT OF CELL GROWTH IN SUSPENSION CULTURES

The cells in suspension culture grow by cell division and the number of cells increases. Growth studies of this kind are very valuable for the characterization of cell lines, effect of nutrient medium and hormones etc.

- **Cell counting:** The rate of increase in cell number can be calculated simply by counting the cell number in haemocytometer under the microscope. Cell count data obtained from haemocytometer is multiplied by a factor X 10^4 and the result can be expressed in terms of cell number per unit volume of culture. Therefore, the growth can be measured by comparing the cell number at the beginning of the culture and after certain days of incubation.

  Again, as the cells increase in number during growth, the liquid medium will be more turbid, and as a result, the OD of the suspension culture will also be altered. The change of OD value can be detected by a colorimeter. Therefore, the growth can be measured from OD value.

- **The PCV:** Definite volumes of the cell suspension can be harvested from multiple replicated sets of culture. Such amount of cell suspension is transferred in graduated conical centrifuge tube and is centrifuged at 1,000 x g for 5 min. The cells form a pellet after centrifugation. The volume of cell pellet then represents the PCV. It is also called biomass volume. Therefore, the growth can be monitored by harvesting the cell suspension at definite periods of interval and then measuring the PCV. This growth can be expressed as millilitre cell pellet per millilitre culture.

- **Cell fresh-weight:** The cells are collected on a preweighed (wet) circular filter of nylon fabric supported in a Hartley funnel, washed with distilled water under vacuum and the filter discs reweighed along with the cells.

- **Cell dry-weight:** After centrifugation, pellet of the cells is obtained. Dry weight of the cell mass can be estimated by drying the pellet in a hot air oven (12 h at 60°C) after replacing the supernatant and then weighing the dried cell mass in a chemical or electrical balance. In this method, growth can be expressed in terms of dry weight in gram or milligram per unit volume of the culture.

5.9. SYNCHRONIZATION OF SUSPENSION CELL CULTURES

Synchronization would require all the cells to proceed simultaneously through specific targets of the cell cycle. A synchronized culture is defined as a culture in which the cell cycles of a proportion of cells are synchronous i.e., majority of the cells proceed through each cell cycle phase simultaneously. In suspension cultures, plant cells vary greatly in size and shape, nuclear volume and DNA content, and cell cycle time. This variation makes the culture asynchronous that complicates the studies of biochemistry, genetic and physiological aspects of cell division. The major obstacles in a suspension culture are low percentage of actively-dividing cells and the tendency of the cells to form aggregates. In order to avoid this and achieve synchronization, various methods have been employed:

- **Cold treatment:** Temperature shocks at 4°C for 4 d can be used.
- **Use of inhibitors:** FudR, excess tritiated thymidine (TdR) and HU are used that temporarily block the progression of events in the cell cycle and accumulate the cells in specific stage. Upon the release, blocked cells will synchronously enter next stage. However, in this method, cell synchronization is limited to any one cell cycle.
- **Starvation method:** Deprivation of an essential growth compound from suspension cultures leads to a stationary growth phase. Resupplying the missing compound or subculturing to a fresh complete medium will allow the growth to resume and may result in synchronization of the cell growth.
- **Colchicine method:** 0.02% Colchicine (W/V) is supplemented to the suspension culture in exponential growth.

5.10. TECHNIQUE FOR SINGLE CELL CULTURE

Single cells can also be isolated from suspension cultures by suitably filtering out cell clumps and harvesting the cells by centrifugation. Single cells can be cultured using the following techniques:

5.10.1. Bergmann Cell-Plating Efficiency Technique

Bergmann first used the most popular technique of plating out cell suspension on agar plates (Fig. 13). The basic technique of plating is to first count the cell number without maceration stage. Both the suspension and nutrient agar media are prepared in double concentration separately. These equal volumes of the suspension and agar medium cooled at 35°C are mixed, and dispensed rapidly into petri dishes in such a manner that the cells are evenly distributed in a thin layer (approximately 1 mm). The suspension culture, which carries cell aggregates, should be filtered through a sieve and only the fine suspension should be plated. After incubation, the number of colonies per plate is counted by making shadow graph printing of each plate under photographic enlarger. When the macroscopic colonies are found, they are isolated and cultured separately.

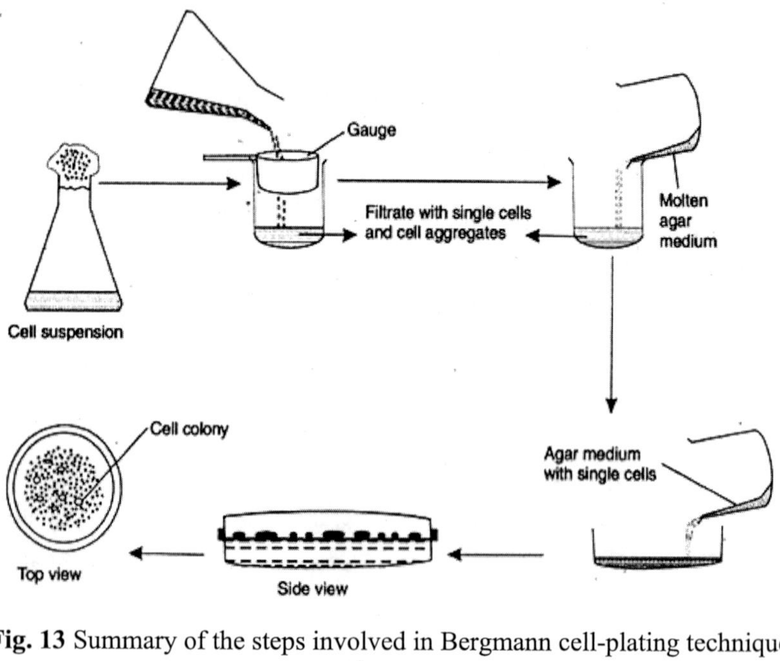

Fig. 13 Summary of the steps involved in Bergmann cell-plating technique

5.10.2. Filter Paper Raft-Nurse Tissue Technique

Single cells are placed on small pieces of sterilized filter paper, which themselves are placed on top of the established callus cultures several days in advance. This allows the filter papers to be wetted by the exudates from the callus tissues. The cells divide and form macroscopic colonies on the filters; the colonies are then isolated and cultured.

5.10.3. Microchamber Technique

A microchamber can be created by using a microscope slide and coverslips or by a cavity slide. Single cells are suspended in microchamber, which is covered by coverslip. In case of cavity slide, the drop is placed on to a coverslip, which is then inverted on to the slide cavity. Microchambers allow microscopic observation and can be kept in a petri dish for incubation.

5.10.4. Microdrop Method

A specially-designed dish, Cuprak dish, having a smaller outer chamber and a large chamber is employed. Microdrops of 0.25–0.5 ml are distributed in the microwells and the dish is sealed with parafilm.

5.11. TESTS FOR VIABILITY OF CULTURED CELLS

The growth of the cultures is largely dependent on the viability of the cells, which can be assessed by microscopic examination of untreated cells or after staining them with substitute chemicals. These include:

- **Phase contrast microscopy:** Cytoplasmic streaming and the presence of a healthy nucleus indicate that the cells are viable. Phase contrast microscopy is recommended as it is difficult to observe these aspects in unstained cells under a bright light.
- **Reduction of tetrazolium salts:** This test is used to measure the respiratory efficiency of the cells by reduction of TTC to the red dye formazon. Formazon can be extracted and measured spectrophotometrically.
- **The FDA method:** The FDA method offers a quick visual assessment of the viability of the cells. Stock solution of FDA (0.5% w/v) is prepared in acetone and stored at 0°C. Viability is tested by adding this solution to the cell or protoplast suspension at a final concentration of 0.01%. Appropriate osmotic stabilizer is added to the FDA solution for protoplasts. After 5 min incubation, the cells are examined under a microscope with a suitable excitation or suppression filter. Though FDA is non-fluorescing and non-polar, it is cleaved by esterase activity inside the living cell, resulting in the release of the polar portion of fluorescein, which fluoresces under UV. Since fluorescein is not freely permeable across the plasma membrane, it accumulates mainly in the cytoplasm of intact

cells, and thus, becomes distinctly visible. In a dead cell, the fluorescein is lost and remains invisible. Fluorescein gives a green fluorescence under UV light.

- **Evan's blue staining:** A dilute solution (0.025% w/v) of Evan's blue dye stains the dead or damaged cells while the living or viable cells repel the dye and remain unstained.

5.12. ADVANTAGES OF CELL AND SUSPENSION CULTURE

- The cell suspension culture systems could be used for large-scale culturing of the plant cells from which secondary metabolites could be extracted.
- This method can ultimately provide a continuous, reliable source of natural products.
- The possible use of plant cell cultures for the specific biotransformation of natural compounds have been demonstrated (Cheetham, 1995; Dixon *et al.,* 2005; Elio *et al.,* 2004). Due to these advances, research in the area of tissue culture technology for production of plant chemicals has bloomed beyond expectations.
- Large-scale plant tissue culture have been found to be an attractive alternative approach to the traditional methods of plantation as it offers a controlled supply of biochemicals, independent of plant availability.

5.13. PLANT CELL REACTORS

Mass culture of plant cells *in vitro* has been proposed as a viable alternative for the production of vast arrays of high value, low volume phytochemicals. Therefore, during the past three decades, considerable work has been done to design bioreactors for plant cell culture. A bioreactor is a glass or steel vessel in which organisms are cultured. Ideally, bioreactors are fitted with probes to monitor the pH, temperature and dissolved oxygen in the culture. They also have provisions to sample the cultures, add fresh medium, adjust pH, air supply, mixing of cultures and controlling the temperature, without endangering the aseptic nature of the culture. It, thus, allows closer control and monitoring of the culture conditions than is possible using shake cultures.

5.13.1. Types of Bioreactors

The major types of bioreactors currently in use for the suspension culture of plant cells are stirred-tank, bubble column, air-lift and rotating-drum reactors.

Stirred-tank reactor (Fig. 14)

In this type, air is dispersed by mechanical agitation. Temperature, pH, amount of dissolved oxygen and nutrient concentration can be controlled better within this reactor than any other reactor. A major drawback of the stirred-tank reactor is the shearing stress generated by its stirring device to which plant cells may be sensitive. Despite this, most existing laboratory and commercial bioreactors are of the stirred-tank design, which have been suitably modified for plant cell culture. Impeller designs for stirred-tank reactor are given in Fig. 15. The two largest reactors (75,000 and 50,000 l) used to-date for plant cell culture are stirred-tank type. Stirred-tank reactors of 200 l and 500 l capacities were used for the first industrial plant cell culture process in which shikonin was produced from *Lithospermum erythrorhizon*.

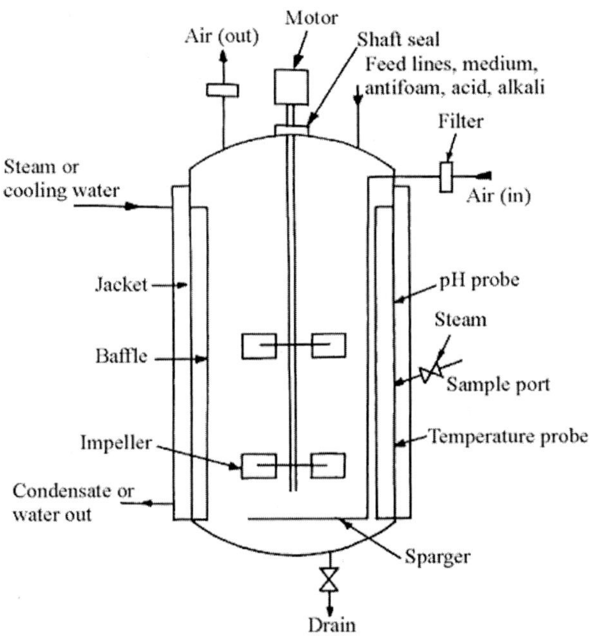

Fig. 14 A stirred-tank bioreactor

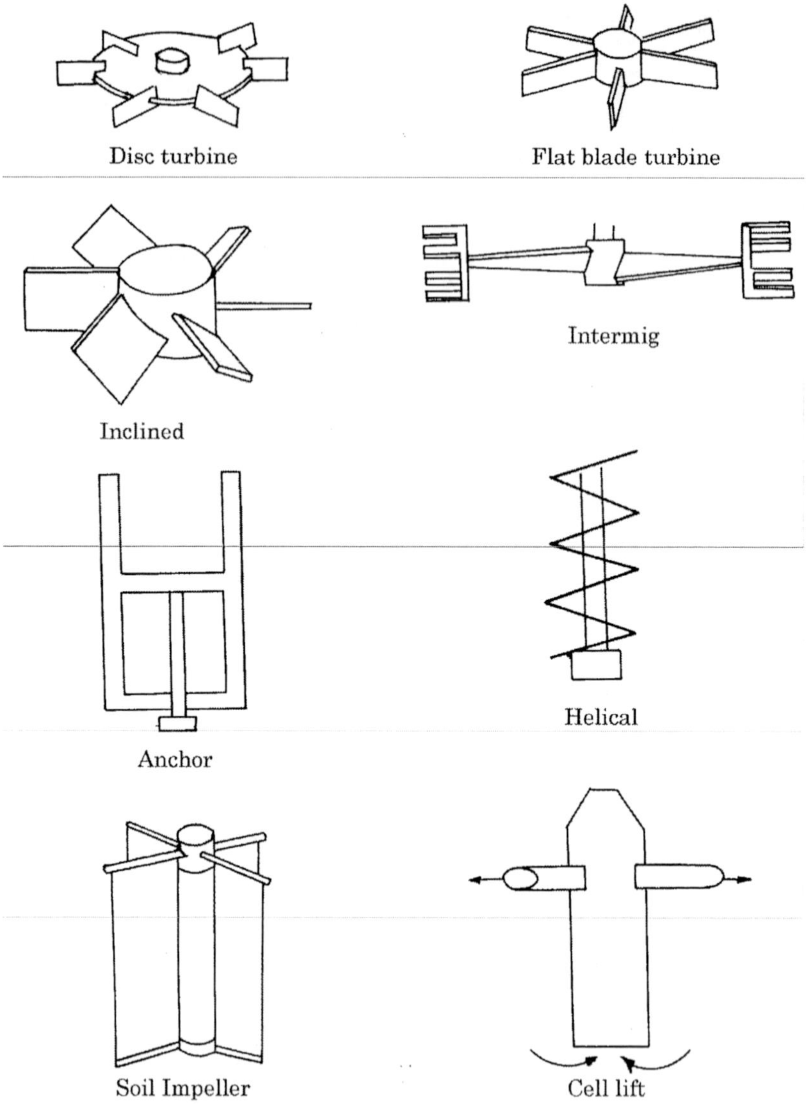

Fig. 15 Impeller designs for stirred-tank reactor

Bubble-column reactor

The bubble-column reactor (Fig. 16) is one of the simplest types of gas-liquid bioreactors used for aerobic fermentation. It consists of cylindrical vessel aerated at the bottom. In such a system, the gas is dispersed pneumatically through a deep pool of liquid by means of nozzles or perforated plates. A 1.5 kl bubble-column reactor has been used for the cultivation of *Nicotiana tabacum*, but insufficient mixing at such a scale reduces the specific growth rate of the cells.

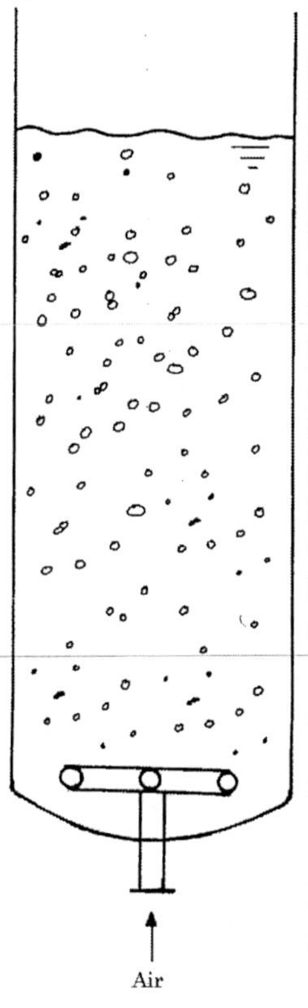

Fig. 16 A bubble-column bioreactor

Air-lift reactor

In air-lift reactor, as its name implies, compressed air is used for aeration and mixing of the contents of the reactor vessel. Its operation is based on the draught tube principle. Air sparged onto the base of the reactor lowers the density of the medium, which rises up to the draft tube, pulling fresh medium in at the base and, therefore, a flow is achieved. Schematic diagrams of the draught tube (inner loop and outer loop) air-lift vessels are given in Fig. 17. A more uniform flow pattern is achieved in the air-lift reactor compared with the bubble-column reactor, where a random flow pattern exists.

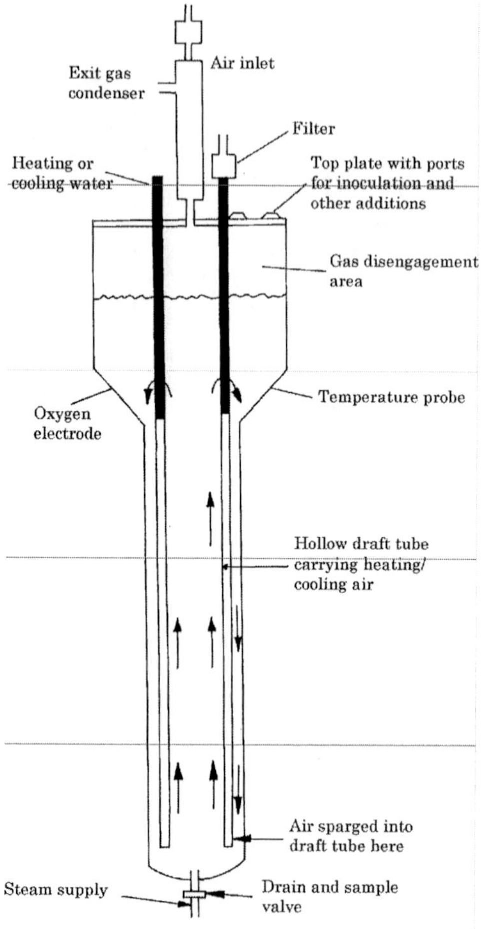

Fig. 17 An air-lift bioreactor

Air-lift reactors up to 100 l capacity have been used extensively by Fowler and his co-workers for cultivation of *Catharanthus roseus* cells. The air-lift reactor is one of the most suitable bioreactor types for culturing of plant cells on a large scale. It provides reasonable mixing and oxygen transfer at low shear, and less contamination occurs because there are no moving parts and no intrusion of impeller shaft.

Rotating-drum reactor

The rotating-drum reactor consists of a horizontally-rotating drum on the rollers connected to a motor. The rotating motion of the drum facilitates good mixing and aeration without imposing a high shear stress on the cultured cells. Baffles in the inner wall of the drum help to increase the oxygen supply. This type of reactor has the capacity to promote high oxygen transfer to the cells at high density. It has been used to grow cultures of *Catharanthus roseus* and *Lithospermum erythrorhizon* up to 1,000 *l* in volume.

In a comparative study of the performance of rotating-drum and stirred-tank reactors for the cultivation of *Catharanthus roseus,* the former was found to be superior on the basis of increased oxygen transfer at high cell densities (Tanaka *et al.,* 1983). The rotating-drum reactor facilitates better growth and imparts less hydrodynamic stress.

5.14. IMMOBILIZED PLANT CELL REACTORS

Immobilization of the plant cells into a suitable carrier and cultivation of these cells in different types of reactors have been developed as an alternative to free cell culture systems for the production of industrial phytochemicals. Entrapment in natural polymers (alginate, agar, agarose, carrageenan) or synthetic polymers (polyacrylamide), adhesion to reticulate polyurethane foam, and confinement behind semi-permeable membranes have been employed to immobilize plant cells.

Alginate has been most popularly used to immobilize the plant cells. Cell suspension in 4% sodium alginate solution is allowed to fall as 2 mm drops in a beaker containing 0.2 M solution of calcium chloride. Calcium alginate is formed by ion-exchange reaction and the drops harden as beads within 20–30 min. Alginate-entrapped cells can be cultured in packed-bed, fluidized-bed or air-lift bioreactors.

Polyurethane foam has been used to immobilize a range of cell lines. The cells are immobilized in these matrices by flowing cells and medium through the foam or by adding sterile foam to the growing cultures. The foam can be cut into various shapes. Polyurethane-entrapped cells have been cultured in both packed and fluidized beds as cubes, shaped into draft tube or threaded as strips on stainless steel rods.

Membrane reactors (e.g., hollow-fibre units and flat-membrane reactors), in which cells are separated from the growth medium by the membrane, are particularly suitable for fragile cells that can be entrapped more readily on the membrane and allow better control over cell density. The environment in a membrane reactor is more homogeneous; pressure drop and fluid dynamics are more easily controlled, and are relatively independent of the scale of operation. In a hollow-fibre reactor, the cells are introduced into the shell side of the hollow-fibre catridge, and the medium is circulated through the fibre lumen and aerated using a separate reservoir. Since the cells do not stick to the fibre membrane, the reactors may retain their mechanical integrity for a longer period of time and may be reusable. When the cells are no longer productive, when an experiment is over, or when a new cell-product combination is desired, it is potentially possible to flush out the old cells and refill the device with the new cells.

In flat-plate membrane reactor system, the cells are loaded manually into the membrane cell layer and direct sampling can be achieved through a removable cap plate. The substrate enters the cell layer by diffusion or pressure-driven flow and is converted into product, which diffuses into the cell-free compartment. Multimembrane reactors have also been proposed for immobilized plant cell cultures. The main advantage of this reactor is that the desired metabolites are produced and selectively separated from the reactant simultaneously.

5.15. CULTURE OF ISOLATED SINGLE CELLS

Free cells isolated either from plant organs (mesophyll tissue) or cell suspensions are grown as single cells under *in vitro* conditions using a suitable medium. This process, called plating is of particular importance when attempting to obtain single-cell clones. Success in the culture of single cells, therefore, depends on the technique and various factors affecting cell plating.

Plating technique

The technique developed by Bergmann (1960) is the most popular one for plating of single cells. Free cells are suspended in liquid medium at a density twice the finally-desired plating cell density. Equal volumes of the media containing single cells and a melted (30–35°C) agar medium (0.6–1%, w/v) are mixed, and rapidly spread out in petri dishes in such a manner that the cells become fixed in an evenly-distributed thin layer (about 1 mm thick) after the agar has cooled and solidified. The dishes are sealed with parafilm and incubated in the dark or diffused light at 25°C. Free cells can also be plated in the liquid medium, but follow-up of the individual cells or their derivatives is difficult in this procedure because the cells do not remain in a fixed position. It is important to note that the suspension cultures are filtered aseptically through a sieve that allows only the single cells required for the culture to pass through; the cell aggregates are discarded.

Plating efficiency

The plates or culture dishes may be observed under an inverted microscope and single cells marked on the outside of the plate with a fine marker to keep track of their regeneration potential. This ensures the isolation of pure single-cell clones. In preparing cultures, if a known volume of suspension is transferred to each plating dish, it should be possible to assess the PE quantitatively using the following formula:

$$PE = \frac{\text{Final number of colonies/plate}}{\text{Initial number of cell units/plate}} \times 100$$

Usually, plating at cell densities of 10^3–10^5 cells ml^{-1} or more yields a high plating efficiency.

■■■■

6 SOMATIC EMBRYOGENESIS

6.1. INTRODUCTION

The act of fertilization triggers the egg cell to divide and develop into an embryo. The process of embryo development is called embryogenesis. However, fertilization is not always essential to stimulate the egg to undergo embryogenesis. As in parthenogenesis, the pollination stimulus alone or simply the application of some growth regulators may induce the egg to undergo embryogenic development. Moreover, it is not the monopoly of the egg to form an embryo. Any cell of the female gametophyte (embryo sac) or even that of the sporophytic tissues around the embryo sac may give rise to an embryo. In several species of *Citrus* and *Mangifera,* the development of adventive embryos (embryos arising from any sporophytic tissue) from nucellar cells is a normal feature.

During the past few years, every plant organ has been shown to form embryos. The whole complex sexual apparatus is not essential for the removal of the effects of ageing and re-establishing of embryonic properties. Embryos formed in the culture have been variously designated as accessory embryos, adventive embryos, embryoids and supernumerary embryos; of these, the term embryoids has been used most widely.

Kohlenbach (1978) has proposed the following classification of the embryos:

- **Zygotic embryos:** Those formed by fertilized egg, or the zyogote
- **Non-zygotic embryos:** Those formed by cells other than the zygote

- Somatic embryos: Those formed by the sporophytic cells (except zygote) either *in vitro* or *in vivo*. Somatic embryos arising directly from other embryos or organs are termed adventive embryos e.g., stem embryos in carrot and buttercup.
- **Parthenogenetic embryos:** Those formed by the unfertilized egg
- **Androgenetic embryos:** Those formed by the male gametophyte (microspores or pollen grains)

Somatic embryogenesis is the process of a single cell or a group of cells initiating the developmental pathway that leads to the reproducible regeneration of non-zygotic embryos capable of germinating to form the complete plants. Totipotent embryogenic cells have been most commonly obtained from the explants of embryonic or young seedling tissues. Excised small tissues from the young inflorescences (before maturation of the floral primordia) are equally effective for the induction of somatic embryogenesis in cultures. Other explants used are the scutellum, young roots, petioles, immature leaf and immature hypocotyls.

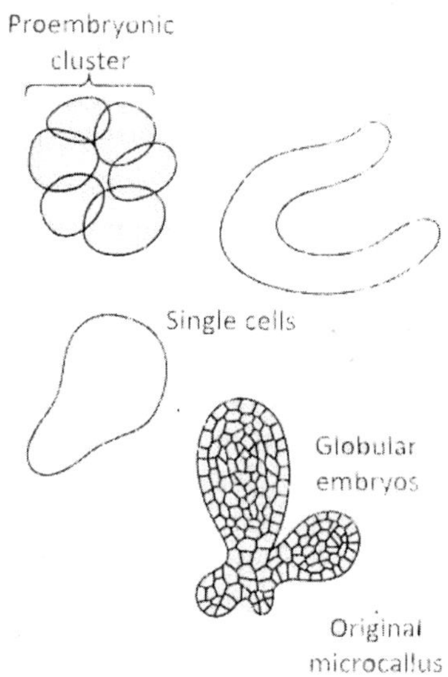

Fig. 18 The embryogenic cells divide in the liquid medium and constitute proembryonic clusters. A certain number of new embryogenic centres will give rise to globular embryos.

In some cultures, somatic embryogenesis occurs directly in the absence of any callus production from PEDCs that are already programmed for embryo differentiation. In the second type of somatic embryo development, prior callus is formed and embryoids originate from IEDCs within the callus tissue (Fig. 18).

6.2 SOME EXAMPLES OF SOMATIC EMBRYOGENESIS

The first observations of *in vitro* somatic embryogenesis were made in *Daucus carota* (Reinert 1958, Steward *et al.,* 1958). Other plants in which this phenomenon has been studied are: *Ranunculus scleratus, Macleaya cordata, Citrus* sps., *Coffea* sps. etc.

In *Ranunculus,* various floral parts (including anthers) as well as somatic tissues proliferate to form a callus on a medium containing CM (10%) with or without IAA. Numerous embryos appear on the callus within 3 wk. The embryos originate from the peripheral as well as deep-seated cells of the callus. Embryo differentiation also occurs in suspension culture raised from these calli. The somatic embryos germinate *in situ* or when they are excised and planted individually on a fresh semi-solid medium. Direct embryogenesis from intact epidermal cells of *Ranunculus* also occurs in the cultures of hypocotyledonary segments and their superficial peels in carrot.

On a medium containing 2,4-D and KN at an equimolar concentration of 5 μmol/1, the individual mesophyll cells of *Macleaya* divided to form tissues, which could be continuously multiplied on this medium. The embryogenic callus selected on the medium containing 0.1 μmol/1 NAA and 1 μmol/1 KN formed embryos at a high frequency upon transfer to a BM that is completely lacking the growth hormones. The embryogenic potential of the callus could be further improved by enriching the medium with reduced nitrogen.

In several species of *Citrus,* the nucellar tissue forms 1–40 adventive embryos per seed, some of which many attain maturity and form plantlets following seed germination. Nucellar polyembryony is of great significance in horticulture for obtaining vigorous seedlings identical to the parent. The embryogenic potential of *Citrus* nucellar cells is also maintained in the cultures. The embryos may arise directly from the nucellar cells or from the callus derived from them. Nucellus only from the micropylar end of the ovule is capable of forming embryos in cultures as it happens in nature. Embryogenic callus could also be raised from single cells, and the protoplasts enzymatically isolated from the nucellar callus of *Citrus sinensis.*

6.3. THEORIES ON EMBRYOGENESIS

Several theories have been proposed to explain the phenomenon of somatic embryogenesis; of which the following are considered important.

6.3.1. Cell Isolation Theory

Steward and his co-workers proposed this theory in 1964. According to them, the embryo-producing cells are isolated from the neighbouring cells in a cell mass. The isolation of cells favours the embryogenesis. The isolation of cell may be induced by the constraints in the surrounding cells due to physical and physiological separation of the cells. In most cases, the connection of plasmodesmata was severed. However, this generally appears to be secondary to the induction process.

6.3.2. Differentiation Theory

This theory states that the embryos would not be produced from the differentiated cells of the explants. The cells of the explants have to undergo dedifferentiation to form a callus. Then the cells of the callus will produce embryos. In other words, dedifferentiation in cells is a prerequisite for the production of somatic embryos *in vitro*. The embryos can be formed directly from the epidermal cells of the stem or hypocotyls. This indicates the possibility of an embryo formation without dedifferentiation. The need for differentiation depends on the explant material used during primary culture. Epidermal cells of the stem, hypocotyl and young embryos may begin embryo development without going through a callus stage while cortical cells, and cells of xylem and phloem explants require dedifferentiation. This theory was proposed by Halperin in 1970.

6.3.3. Intercellular Communication and Cytodifferentiation Theory

According to this theory, cytodifferentiation in the cells due to intercellular communication induces embryo formation. The cytodifferentiation is regulated by diffusion gradients of nutrients, endogenous PGRs and gaseous factors such as oxygen, carbon dioxide and ethylene. The changing microclimate in the culture environment affects intercellular communication and, in turn, cytodifferentiation. This concept was proposed by Street (1973).

6.3.4. Explant Physiology and Culture Environment Theory

This concept was developed by Street in 1976. He is of the view that embryogenesis is a phenomenon, which is dependent on the explant and the culture environment. Explants such as flower buds, young embryos and parts of young seedling are most responsive to produce somatic embryos, but not from those of the mature plants. Apart from the explant physiology, culture environment is also a factor influencing the embryogenesis. For example, highly-embryogenic callus culture can be maintained as non-embryogenic if the medium is supplemented with high level of auxin, and the same may be induced to produce embryos when transferred to an auxin-free medium.

6.3.5. Predetermination Theory

This was proposed by Tisserat *et al.* (1979). It states that the embryo-production potential is predetermined phenomenon in the cells, and the *in vitro* culture provides the opportunity for embryogenesis. In other words, embryogenesis from a cell is an inherent one, which is facilitated to produce embryos by optimal culture environment.

6.3.6. The PEDC and IEDC Theory

Though the embryogenesis is predetermined one, there are instances of non-formation of embryos directly from the explants. In these cases, an intervening callus stage comes between the primary explant and the embryos. The cells in the calli are induced to produce embryos by the manipulation of the medium with relevant growth regulator. Based on this, the above theory was proposed by Sharp and his co-workers. According to this theory, there are two types of embryogenic cells: PEDCs and IEDCs. In PEDCs, embryogeny is determined prior to mitosis while in IEDCs, the embryogeny is induced by providing suitable mitogenic substance i.e., the embryogeny is induced in the cells of the callus by the application of PGRs. Thus, in the callus, embryogenic precursor cells or embryogenic mother cells are formed, which then develop into embryogenic cells. Later, these cells undergo polarized cell divisions typical of normal embryogenesis by forming globular, heart and torpedo-shaped embryos.

6.4. FACTORS AFFECTING SOMATIC EMBRYOGENESIS

Steward *et al.* (1964) emphasized the importance of CM for *in vitro* somatic embryogenesis in carrot, but this could not be confirmed by others. The work to-date has highlighted the importance of two media constituents in somatic embryogenesis *viz.*, auxin and source of nitrogen.

1. **Auxin:** The presence of auxin in the medium is generally essential for embryo initiation. Tissues or calli maintained continuously in an auxin-free medium, generally, do not form embryos; therefore, somatic embryogenesis is achieved in two steps:

    a. First, the callus is initiated and multiplied in a medium rich in auxin (2,4-D 0.5 mg/l), which induces the differentiation of localized groups of meristematic cells called ECs.

    b. Second, the ECs develop into mature embryos when transferred to medium with very low level of auxin (0.01–0.1 mg/l).

The ECs continue to multiply without the appearance of mature embryos in repeated subcultures on the proliferation medium. The presence of an auxin in the proliferation medium seems essential for the tissue to develop embryos in the embryo development medium. The tissue maintained continuously in an auxin-free medium would not form embryos. In this respect, the proliferation medium could be regarded as the induction medium for somatic embryogenesis and each EC as a disorganized embryo.

In *Coffea arabica* also, somatic embryos develop only when a callus grown on 2,4-D-containing medium is transferred to 2,4-D-free medium. In pumpkin, NAA and IBA favoured embryogenesis.

Wochok and Wetherell (1971) have suggested that the 2,4-D-induced suppression of embryo maturation may be mediated through endogenous ethylene production. High ethylene content would result in an enhanced activity of cellulase or pectinase or both, causing breakdown of the clumps before polarity is established in proembryos for further organized development. Thus, tissue multiplication goes on in 2,4-D medium, but the mature embryos do not appear.

2. **Cytokinins:** In carrot (*Daucus carota*), both BAP and KN were inhibitory for embryogenesis, but zeatin (0.1 μmol/1) promoted the process. However, callus cultures from the leaves of *Coffea arabica* required a high kinetin to auxin ratio for a high-frequency embryo induction throughout the culture period. The BAP in the proliferation medium may promote cell division, but usually it inhibits the embryogenic potential of many plants. This may be due to selective multiplication of non-embryogenic cell components of the cultures.

3. **Gibberellin:** Gibberellin inhibits somatic embryogenesis in almost all the species. In combination with zeatin, GA_3 produces an increased frequency of abnormal embryo development.

4. **Abscissic acid:** The ABA often positively influences the maintenance of somatic embryogenesis and reduces the frequency of abnormalities found in developing embryos. The IAA, ABA and GA_3 have been reported to suppress embryogenesis in carrot, *Citrus* etc.

A micropropagation procedure for *Anthurium scherzerianum* has been developed using plantlet regeneration via adventitious shoot-bud induction from leaf or spadix derived (caulogenic) callus on Nitsch medium with low ammonium nitrate content and supplemented by 1 mg/l BAP and 0.1 mg/l 2,4-D (Geier, 1988). The plantlets were stable and there was evidence that organized shoots exerted a stabilizing influence on the ploidy status of the surrounding tissue. Pierik and co-workers (1973), and Kunisaki (1980) have been successful in obtaining regeneration via callus and by shoot-tip culture of *Anthuriums*.

An improved method for rapid callus proliferation and regeneration of *Fuchsia hybrida* on a BM containing 5×10^{-6} M IAA and 5×10^{-6} mM BAP have been established. Regeneration could be secured on transfer to a medium at half strength (Chow *et al.,* 1990).

In vitro propagation and corm formation in some commercial cultivars of *Gladiolus* have been initiated from axillary buds excised from cold-stored corms cultured on MS medium + BAP (0.5 mg/l). Proliferation of the shoot buds occurred in 4–5 wk after inoculation (Dantu and Bhojwani, 1988). Shoots were induced to form corms *in vitro* to facilitate transplantation. Small storage organs induced *in vitro* may be exploited for the purposes of micropropagation such as the bulbils of *Amaryllis*, lily and others, cormlets of *Gladiolus* and protocorms of orchids. Small corms or cormlets of *Gladiolus* may be formed directly from the tissue explants or even from the cultured callus. Protocorms are special reproductive propagules formed by the orchids – structures developed by the embryo or shoot apical meristems. If the protocorms are vertically divided into two or more parts, each piece could regenerate a plantlet.

In vitro propagation of *Hydrangea quercifolia* has been successful through differentiation of the leaf callus into plantlets and stimulation of axillary bud release from dormant shoots. The regenerated plants are uniform in growth pattern and easy to root (Sebastian and Heuser, 1987). A micropropagation system has been developed to facilitate release of unique pink or white-flowered selections of *Yucca glauca* (Bentz *et al.,* 1988). A rapid and effective method of plant micropropagation by vegetative meristems has been established for *Aloe barbadensis* by *in vitro* culturing of shoot apices on the medium containing the hormones 2,4-D + BA (Natali *et al.,* 1990).

5. **Nitrogen sources:** Besides auxin, the form of nitrogen in the medium significantly affects *in vitro* embryogenesis. Nitrogen in reduced form, such as ammonium salts, is required for both embryo initiation and maturation. The complex additives such as CM, CH, single amino acids or mixture of 18 amino acids can be used as source of nitrogen. Halperin and Wetherell reported that embryo development in the wild carrot callus cultures occurred only if the medium contained some amount of reduced nitrogen. The calli initiated on the medium with potassium nitrate as the sole source of nitrogen failed to form embryos upon removal of auxin. However, the addition of a small amount (5 mM/1) of ammonium chloride (NH_4Cl) in the presence of 55 mM/1 of potassium nitrate allowed the embryo development. If a callus was raised on a medium with potassium nitrate + ammonium chloride, it would form embryos irrespective of the presence or absence of ammonium chloride in the differentiation medium. With an optimal combination of ammonium chloride (10 mM/l) and potassium nitrate (40 mM/l), the number of embryos per 2 ml of the suspension can be as high as 1131 ± 219 as compared to 0.9 ± 0.3 and 11.3 ± 4.1 with potassium nitrate alone at a concentration of 55 mM and 95 mM/l, respectively. The presence of a minimal amount of endogenous ammonium ions (around 5 mM/kg fresh weight of tissue) is essential for the

occurrence of embryogenesis in the cultured cells of carrot. Unless this threshold level is achieved, the cell cannot embark upon embryo development.

6. **Carbohydrates:** Sucrose reduces somatic embryogenesis, but other sources such as fructose, dextrose and mannose have been used for advantageous purpose.

7. **Other factors:** Genotype of the explants has a marked influence on somatic embryo regeneration and it has been shown in many cases such as alfalfa, wheat, maize etc. There has been little work in the effect of temperature, especially cold treatments, on somatic embryogenesis. Light intensities also affect the process of embryogenesis. Brown *et al.* (1976) reported that high potassium (920 mM/l) is necessary for embryogenesis in wild carrot. The amount of dissolved oxygen in the medium should be below the critical level of 1.5 mg/l to allow embryo development. Higher levels of dissolved oxygen favour rooting. The AC is also reported to improve embryogenesis in carrot. It has been suggested that charcoal may absorb a wide variety of inhibitory substances as well as hormone.

6.5. ANATOMY AND CYTOLOGY TO SOMATIC EMBRYO DEVELOPMENT

Steward *et al.* (1964) suggested that the embryos are originated from single cells. They also said that the isolation of a cell from its neighbouring tissue is necessary for it to develop into an embryo. However, detailed studies on *in vitro* somatic embryogenesis revealed that the cells remain in cytoplasmic continuity with the adjacent cells during the induction and early stages of embryogeny. Isolation occurs at a later, multicellular stage.

Anatomical and histological studies of callus initiation and tissue growth in the suspension cultures of carrot have revealed that a culture carried two types of cells;

1. Large and highly-vacuolated cells, normally lacking embryogenic potential that are freely dispersed in the medium

2. Fairly-small and densely-cytoplasmic cells capable of forming embryos, which occur in clusters or clumps. The cell clusters capable of embryo formation have been termed ‘proembryonic

masses' by Halperin (1966) and ECs by the cutinization of its outer surface. However, the cell clusters are released from the clumps at various stages of development, generally, after the globular stage. Their further development occurs as free-floating structures.

6.5.1. Maturation of Somatic Embryos

Often the embryos mature and germinate in the original medium. However, in the species requiring cold treatment for seed germination, it may be necessary to chill the young or mature embryos for their normal development into plants.

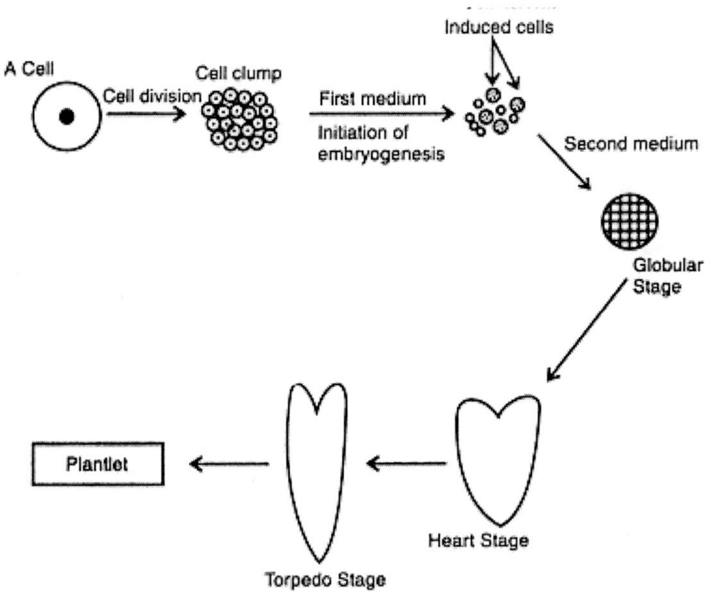

Fig. 19 Schematic representation of somatic embryogenesis

6.5.2. Procedure for Somatic Embryogenesis

1. Use leaf petiole (0.5–1 cm) or root segments from seven-day old seedlings (1 cm) or cambium tissue (0.5 cm^3) from storage root as explant. Obtain leaf petiole and root segment from aseptically-grown seedlings. Obtain cambium tissue from surface-sterilized storage tap root.

2. Following aseptic technique, place the explants individually on a semi-solid MS medium containing 2,4-D and sucrose. Incubate the cultures in the dark. In this medium, the explant will produce sufficient callus tissue.

3. After 4 wk of callus growth, initiate cell suspension culture by transferring the callus tissue to a 250 ml of Erlenmeyer flask containing 20–25 ml of liquid medium of the same composition as used for the callus growth (without agar). Place the flasks on a horizontal gyratory shaker with 125–160 rpm at 25°C. The presence or absence of light is not critical at this stage.

4. Subculture the cell suspensions every 4 wk by transferring 5–65 ml of the fresh liquid medium.

5. Pass the cell suspension through a series of stainless steel mesh sieves to induce a more uniform embryo population. For carrot, the 74 μ sieve produces a fairly-dense suspension of single cell and small multiple clumps. In order to induce somatic embryogenesis, transfer the portions of the sieved cell suspension to 2,4-D-free liquid medium, or plant the cell suspension in semi-solid MS medium devoid of 2,4-D. Incubate the cultures in dark.

6. After 3–4 wk, the culture would contain numerous embryos in different stages of development such as globular stage, heart-shape stage and torpedo stage.

7. Place the somatic embryos on agar medium devoid of 2,4-D for plantlet development.

8. After attaining the torpedo stage, collect the embryoids and transfer them to an enriched MS medium containing high levels of sucrose, nitrogen and sulphur. During this step, the embryoids accumulate food reserves, and increase in size and dry weight. This is often called first maturation phase.

9. Add ABA to the culture medium for the normal embryo development and to inhibit precocious germination, especially root elongation. After reaching 1–2 mg dry weight, transfer the embryoids to a modified medium containing ABA for 3 d. Now, the embryoids grow in size and become resistant to desiccation. This is called second maturation phase.

10. Collect the embryoids from the culture and wash them with sterile water to remove nutrients and sugars. Allow these to dry and store them in a sealed chamber till the moisture becomes 15%.
11. Otherwise, transfer the embryoids to regeneration medium to produce plantlets.
12. Finally, transfer the plantlets to Jiffy pots or vermiculite for subsequent development.

6.5.3. Advantages of Somatic Embryogenesis

- Large numbers (e.g., 60,000 embryos/l of the medium) of the embryos are produced from a small bit of callus.
- The presence of both root and shoot meristems in the same unit
- Easy scale up
- Transfers and other manipulations since embryos can be grown individually and freely-floating in the liquid medium.
- The promise of dormancy induction and long-term storage
- The possibilities of encapsulation and other methods of packaging, and of direct delivery systems as naked embryos or young plantlets

6.5.4. Applications of Somatic Embryogenesis

- **Clonal propagation:** Embryos formed directly from PEDCs produce relatively-uniform clonal material.
- **Somaclonal variants in tree species:** Embryos formed indirectly involving IEDCs generate a high frequency of somaclonal variants. Mutation during adventive embryogenesis may give rise to mutant embryo, which on germination would form a new strain or a plant.
- The large-scale production of somatic embryos can be done by the usage of different types of bioreactors i.e., rotating-drum bioreactor, gaseous-phase bioreactor, light-introducing bioreactor, air-lift column bioreactor etc. e.g., Alfalfa (McDonald and Jackman, 1989).

- As somatic embryos remain virtually unprotected and more or less give rise to genetically-uniform plants, it is easier to study genetic transformation in tree species. They are competent to express the introduced DNA and regeneration of non-chimeric plants. Some notable examples are *Juglans regia* (McGranahan 1990) and *Mangifera indica* (Mathews *et al.*, 1992).
- Plants derived from nucellus and shoot tips are free of virus, and can be used for raising virus-free clones. For example, *Citrus* plant propagation from embryogenic callus of nucellar origin is free of virus.
- Synthetic seeds consisting of somatic embryos encapsulated in a protective coating have been proposed as a low-cost high-volume propagation system.
- Storing of the somatic embryos in LN has attracted considerable attention, especially for the conservation of germplasm.

7 SYNTHETIC SEEDS

7.1. INTRODUCTION

The concept of 'synthetic seeds' was first introduced by Toshio Murashige in 1977, and later became a reality for use by Kitto and Janick (1982) and Redenbaugh and co-workers (1984). 'Synseed' is referred in a narrow sense to 'encapsulated somatic embryo, which functionally mimics the seeds and can develop into seedlings under suitable conditions'. In broader sense, it also refers to encapsulated buds, bulbs, or any form of meristems, which can develop into plantlets. The main thrust idea is to prepare a simple inexpensive delivery unit of tissue culture propagated plants.

The first successful example of synthetic seed technology was in alfalfa and celery. In the case of alfalfa, synthetic seed method was used, not for propagation, but as a tool for breeding and variety development such as hybrid alfalfa. Fuji *et al.* (1989) could grow plants from synthetic seeds of alfalfa planted directly in the field, albeit at low frequency. In the case of celery, Kirin Brewery Co. Ltd., Japan and Plant Genetics Inc., USA conducted the field trials in Japan with 20,000 F_1 synthetic seeds in 1988.

7.2. ENCAPSULATION

For encapsulation process, it is often a practice to coat the propagules in a matrix, which may serve as synthetic endosperm containing carbon sources, nutrients, growth regulators, antimicrobial agents etc. The coat should be non-toxic to the propagule, should protect it from mechanical damage during handling, and allows the development and conversion to occur without any undesirable variation. Several agents including sodium alginate with gelatin, potassium alginate, guar gum, agar, tragacanth gum, sodium pectate, carboxymethyl cellulose, carrageenan with locust bean gum, gelrite and polyco2133 etc. are used for encapsulation (Redenbaugh *et al.,* 1986, 1987), and among these, sodium alginate has been extensively used. In addition, polyethylene oxide homopolymer, acrylic copolymer containing carboxyl groups, potassium propenoate-propenamide copolymer, carboxyl methylized cold-soluble swelling starch, synthetic trioctahedral smectite, synthetic sodium-magnesium-lithium silicate, starch plus synthetic polymer or acrylamide and sodium acrylate, anionic flocculant have also been tried as coating agents (Janick *et al.,* 1993).

There are two types of synthetic seeds: (1) Hydrated and (2) Desiccated.

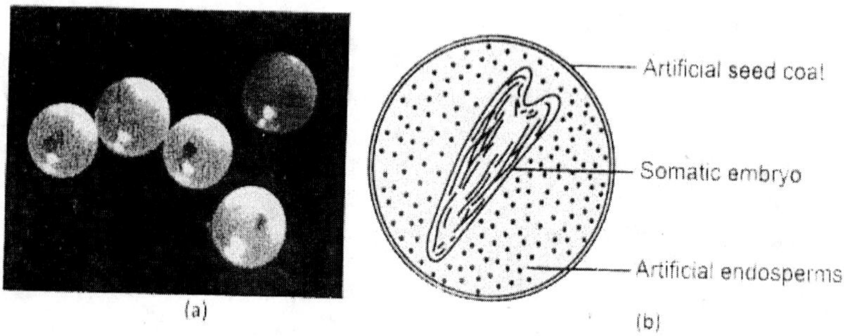

Fig. 20 (a) Photograph of synthetic seed and (b) Capsule gel with hydrophobic membrane.

1. Hydrated synthetic seeds consist of somatic embryos individually encapsulated in hydrogel such as calcium alginate (Fig. 20). Hydrated synthetic seeds are prepared in the following manner. Somatic embryos are mixed with 2% (w/v) solution of sodium alginate, which is dropped by using a plastic pipette into a 100 mM solution of calcium chloride. Surface complexing begins immediately and the gelling is complete within 30 min. The calcium alginate capsule is generally wet and also sticky causing problems for handling. Redebaugh *et al.* (1987) tested various hydrophobic coatings and found Elvax polymer coating to be very effective for yielding 80% shoot emergence. Alternatively, the embryos could be mixed in a temperature-dependent gel such as gelrite, placed in the well of microtitre plate, and gelled as the

temperature was lowered e.g., celery, carrot, cotton, lettuce, corn, alfalfa, sandalwood etc.

2. Desiccated synthetic seeds are produced by coating a mixture of somatic embryos in polyoxyethylene glycol. In this method, the coated mixture is allowed to dry for several hours on a Teflon surface in a sterile hood. The dried mixture is then hydrated and placed on a medium for plantlet development

According to Redenbaugh (1990), both these methods are not completely satisfactory for embryo survival. Hydrated capsules are difficult to store because of requirement for embryo respiration. Also, the capsules dry out unless kept in humid environment or coated with hydrophobic membrane. In the case of desiccated capsules, the process of desiccation itself may damage the embryo.

It varies from species to species whether somatic embryos are to be desiccated or not since in soybean, conversion has been improved from 30–60% in desiccated somatic embryos whereas conversion was reported to be 59% in *Triticum*, which was less than achieved with non-desiccated embryos.

In a few experiments, the artificial seeds were planted in the greenhouse with plant production of 97% for alfalfa and 10% for celery. Sowing is done by fluid drilling of the pregerminated embryos in order to ensure synchronous germination in the field so that the crop may be homogenous.

7.3. PRODUCTION OF ARTIFICIAL SEEDS

The procedure for production of artificial seeds is as follows (Fig. 21):

1. Mix the somatic embryos with a suspension of nutrient medium containing sucrose medium. Inoculation density of 0.8 g of embryo per 25 ml of the medium is suitable for synthetic seed preparation and storage.

2. Drop the alginate gel containing embryos into the calcium chloride solution to form round firm beads. Retain these beads in that solution for 30 min to form a rigid capsule by ion exchange between sodium ions in sodium alginate with calcium ions in calcium chloride solution. Thus, the hardness and rigidity of the capsule depends on the concentration of calcium and sodium ions.

3. Sieve the beads through a nylon mesh to collect the artificial seeds. Reuse the remaining calcium chloride solution to drop fresh somatic embryos.
4. Test the viability by placing them on a moist filter paper in a petri dish or sowing them on a soil treated with fungicides and pesticides.
5. Then store the artificial seeds at 4°C in a dark room in the field.

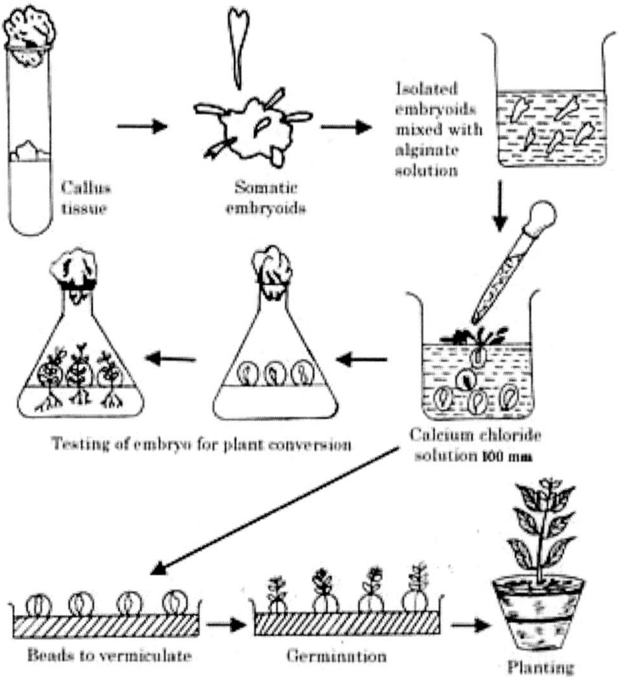

Fig. 21 Flow diagram showing the method for making artificial seeds

One limitation that must be overcome for developing artificial seeds is storage time. Currently, the conversion frequency of encapsulated alfalfa somatic embryos, when stored at 4°C, is high (80%), but embryo germination declines with time. Also, the germination and continued development of the embryos was very inconsistent in different species. Embryo conversion frequency (i.e., embryo to plant development) is the percent of the somatic embryos that produce green plants having a normal phenotype. This assay is critical for developing conditions and media that select for uniformity in crop production. A method to encapsulate somatic embryos is given in amplified and schematic manner in Fig. 21.

Sandalwood plantlets have been propagated from synthetic seed and by encapsulated axillary buds containing fungicide (Bapat and Rao, 1988, 1990).

Table 8: Important crop plants in which artificial seed production and plant conversion has been demonstrated

***In vitro* propagules for encapsulation**	**Crop**
Somatic embryos	Alfalfa, celery, brinjal, carrot, *Brassica*, lettuce, sandalwood, rice, horseradish
Axillary buds/adventitious buds	Mulberry, *Eucalyptus*, *Vitis*, sandalwood
Shoot tips	Banana, caradamon, *Carum carvi*

7.4. APPLICATIONS

Synthetic seed technology offers many useful advantages on a commercial scale for the propagation of a variety of cereal plants, crops for which true seeds are not used or readily available for multiplication (e.g., potato) or the true seeds are expensive (e.g., cucumber and geraniums), hybrid plants (e.g., hybrid rice) and many vegetatively-propagated plants, which are more prone to infections (e.g., day lily, garlic, potato, sugarcane, sweet potato, grape and mango). This technology would also be useful for multiplying genetically-engineered plants (transgenic plants), somatic and cytoplasmic hybrids (obtained through protoplast fusion techniques), sterile and unstable genotypes. Besides, artificial seeds would be useful material for preservation of desirable elite genotypes (cryopreservation). They would also be valuable tools in experimental research to study the process of zygotic embryogenesis and understanding the role of endosperm in normal embryo development and germination.

The artificial seed coating also has the potential to hold and deliver beneficial adjuvants such as growth-promoting rhizobacteria, plant nutrients and growth control agents, and pesticides for precise placement.

The potential uses of artificial seeds are two-fold; as a delivery system and as an analytical tool. Artificial seeds offer the possibility of a low-cost, high-volume propagation system that will compete with true seeds and transplants. Most likely, the technology will first be used with hybrid vegetable crops such as celery or with high-value flower and ornamental species. Because of the relative ease of producing large number of somatic embryos, artificial seeds will be applicable for monoculture as well as mixed genotype methods of agriculture.

7.5. LIMITATIONS

There are also certain limitations, which have to be looked into. The plant should have the capacity to produce large number of propagules (somatic embryos or any other meristems) over a long time span that should develop synchronously, and also withstand the desiccation shock and retain the ability to develop into plantlets.

There is a great need for refinement of this technology by tackling certain technical problems such as the need to produce high-quality and high-fidelity somatic embryos, and to avoid the genetic instability and variability of tissue culture derived plants. Further, the understanding about the storage, transport, handling, and growth habit and harvest index of the artificial seeds is essential.

During the recent past, for many of the important, high-value genotypes, somatic embryogenesis has not been reported. Therefore, there is a need to shift the focus on somatic embryogenesis in valuable crop plants. Research on encapsulation of the axillary buds, shoot tips or any other propagules should be taken up. Equally important is the need to develop new synthetic seed-coatings for encapsulating embryos and other vegetative propagules. Synthetic seed technology, in years to come, would certainly find its application in plant propagation and delivery of the tissue cultured plants.

7.6. USES OF ARTIFICIAL SEEDS

- They are used for the propagation of non-seed producing plants and polyploids, which have high economic value.
- They are used to propagate male or female sterile plants for breeding programs.

- Transgenic plants with original genotype can be stored in the form of somatic embryos in artificial seeds.
- Pathogen-free plants can be produced from artificial seeds. Hence, seeds can be transported easily without much risk of quarantine and bulkiness.
- They are potential tools for genetic engineering in plants.
- Cryopreservation of artificial seeds is used for germplasm storage of plants such as coconut, cocoa etc.

7.7. REFERENCE

Redenbaugh K (ed) 1993 Applications of synthetic seeds to crop improvement, Boca Raton: CRC Press.

■■■■

8 MICROPROPAGATION OF HIGHER PLANTS

8.1. INTRODUCTION

The success of plant tissue culture techniques has made large-scale cloning of the plant species possible. It is possible to produce, literally, millions of high-quality and genetically-uniform plants with a relatively small investment in space, technical support and materials. This process is known as micropropagation. The most common technique is to place an excised meristematic tissue on a medium that reduces apical dominance and encourages axillary bud development. The new shoots can be separated and subcultured to produce more axillary shoots, or can be placed on a medium that encourages rooting, and hence, can be planted out. Alternately, tissues can be used to establish callus cultures, which may be induced to form roots and shoots.

Micropropagation can also be an effective way to eliminate viruses and other pathogens, and produce commercial quantities of pathogen-free propagules. The first plants to be mass-produced by tissue culture were virus-free *Cymbidiums* (Orchids), but the technique has also been found useful for potato, lilies, tulips and other species that are normally propagated vegetatively. Potato, for example, is vegetatively propagated through buds on the tubers; a system that readily transmits the viruses to the next generation. Micropropagation of potato from meristem cultures has proven to be an effective way to isolate virus-free lines.

Micropropagation is now also being used extensively in the production of forest tree species. Here, the propagules are generated primarily from cultures of axillary and adventitious buds; callusing and differentiation of new buds is rarely used. A similar approach has been applied successfully to cultivars of apple (*Malus*), peach (*Prunus*) and pear (*Pyrus*). Since most of the temperate fruits are heterozygous, they do not breed true from seed, but are propagated by vegetative cuttings.

Rooting of microcuttings in culture is now a routine procedure in many commercial laboratories. By 1983, growers in Netherlands were producing more than 21 million plants by micropropagation.

In spite of the fact that plantlets derived from tissue culture are cloned from presumably identical somatic (non-sexual) cells, it is not clear that the regenerated plants exhibiting significant variation, but it involves spontaneous genetic variation as a result of the culture conditions. The value of somaclonal variation is that occasionally, the variants exhibit disease resistance or some other agronomically-useful trait.

Micropropagation is also known as *in vitro* regeneration of the plants or clonal propagation of the plants. The main intention of clonal propagation is to establish the plants that are uniform and predictable of selected qualities. In view of the substantially-superior rate of plant increase, interest in the use of tissue cultures as an alternative to traditional asexual multiplication methods is currently spreading rapidly. The tissue culture approach provides other equally-significant benefits:

- Pathogen-free plants are recoverable from clones that are systematically infected.
- Elite species can be multiplied on a large scale using *in vitro* methods.
- International or regional exchange of genetic stock is expedited and rendered safer.
- Rapid clonal increase also signifies faster and timely establishment of new cultivars.

The application of *in vitro* methods to plant propagation is due to the contributions of late George and Morel, and their colleagues. They first obtained virus-free plants in the year 1952 and later extended the technique to orchid multiplication in 1960. The widespread use of tissue culture has depended on the identification of plant hormones, and particularly on the disclosure of their roles in organogenesis. The relationship of auxin-cytokinin balances is important for root and shoot initiation (Skoog and Miller, 1957). *In vitro* propagation methods are generally used for the multiplication of:

- Herbaceous plants
- Woody genera (forest trees)

- Ferns
- Ornamental plants
- Monocotyledonous plants
- Horticultural plants
- Fruit plants
- Vegetable crops
- Dicots

Clonal propagation aims at the production of plants breeding true-to-type, in larger numbers, in quicker time and disease free. In general, this is accomplished by *in vitro* methods involving meristematic explants, shoot-tip culture, adventitious bud induction or by stimulation of axillary or lateral meristems into their growth and development, or through culture of non-meristematic explants (leaf, petiole, root etc.). By far the most versatile method of ensuring uniformity of the crop is through culture, but rather restrictive in commercial practice for mass propagation unless multiple shoot induction through axillary meristems can be secured. Young leaves, immature embryos, seedling hypocotyl, cotyledons, ovary and ovule, and juvenile tissue of woody plants are good sources of explanted material to initiate tissue cultures for morphogenetic manipulations.

8.2. SOMATIC CELL EMBRYOGENESIS

Somatic embryos are simply organized structures, the morphology of which is fundamentally identical to that of embryos originating in zygotes, but the initials of which are somatic cells. Clonal plant multiplication occurs fastest through somatic cell embryogenesis. Yields of 500 carrot embryos/gram callus/ month are usual. Nevertheless, the method is severely limited in its applicability. The process remains restricted to a few genera, even then, only to the tissues excised from embryos, seedlings and other juvenile materials. Generally, somatic embryogenesis is accomplished in two steps: (1) increase of callus and (2) development of the embryos and plants. Stock of callus is established through monthly subcultures in a nutrient enriched with ammonium nitrate and the auxin, 2,4-D. Transfer of the callus to auxin-free or low-auxin medium results in emergence of the embryos and complete plantlets. Embryo formation frequently occurs directly in the explants. In such cases, the intermediary callus may be avoided, and the plant multiplication can be achieved by separating individual embryos and culturing them further in the fresh medium. New embryos probably arise by budding from existing embryos.

8.3. PRODUCTION OF ADVENTITIOUS SHOOTS AND RELATED ORGANS

Adventitious shoots and related organs are stem and leaf structures that arise in tissues located in the sites other than the normal leaf axil region. The structures in question include the usual stems, bulbs, corms, tubers, rhizomes, and others. The development of these structures *in vitro* has been more probable than that of somatic embryos, although plant multiplication is substantially slower.

The production of adventitous shoots and related organs might also proceed through an intermediary callus. However, the advisable method is to avoid the establishment of callus cultures. Increase of propagule is accomplished by separating and subculturing the emerged organs at 4–6 wk intervals. Organogenesis is stimulated by employing a medium containing moderate amounts of auxin and cytokinin e.g., 2–3 mg/l each of IAA and KN. When the desired number of shoots is attained, they are rooted through stage III. Bulbs, corms, tubers and storage organs require no rooting.

The problem of genetic variability may be, especially pronounced among plants arising through adventitious organogenensis. Repeated reculturing, generally, results in an increasing incidence of deviant plants, and diminishing rate of plant multiplication. Plants, the desirable qualities of which are attributable to chimera, should not be propagated through this method.

8.4. ENHANCED AXILLARY BRANCHING

Axillary branches are simply the shoots that emerge from their normal position i.e., the leaf axil. Lateral buds and stem terminals, containing quiescent or active shoot tips, are stimulated by their axillary branching process by supplying a nutrient medium containing extremely high concentrations i.e., 10–30mg/l of cytokinin. At intervals of 4–6 wk, the miniature branches are severed and subcultured individually in fresh medium. After several repetitions of the subculturing process, the tiny branches are transferred to stage III and rooted. A million-fold enhancement of the clonal plant increase in a year's time over traditional methods has been achieved with some genera e.g., *Gerbera*. The plants obtained by this method are genetically less variable than those arising by other methods.

8.5. STEPS INVOLVED IN PLANT MULTIPLICATION

Micropropagation is now typically divided into 5 stages. Stages I–IV were originally proposed by Murashige. Debergh and Maene added stage 0 (Fig. 22).

1. Stage 0: Donor plant selection and preparation

Explant quality and responsiveness is influenced by the physiological and phytosanitary conditions of the donor plants. The steps in this stage are:

- Donor (stock) plants indexed for pathogens
- Pathogen-free stock plants maintained in clean conditions (low humidity, drip irrigation)
- Vigorous growth is encouraged, but not over fertilization
- Donor plants may be pretreated in certain ways

2. Stage I: Preparation and culture of explants

Steps involved in Stage I are:

a. Elimination of microbial contamination

    i. **Surface contaminants:** Principally, microbial saprophytes that are eliminated by surface disinfection

    ii. **Internal contaminants:** Principally, pathogens that are eliminated by thermotherapy (35–40ºC) and culture of explants free of organisms or by antibiotics

b. **Maintenance of asepsis (free from microorganisms) during excision and culture:** Procedures are carried out in sterile laminar flow positive-pressure hoods (0.3 μm HEPA filters).

The aim in Stage I is to achieve prolonged survival of the explant *in vitro* without any contamination. The nutrient formulation and culture environment of Stage II may satisfy the Stage I requirements. Neverthless, some plants exhibit discolouration of the tissue and/or medium and precocious deterioration of the cultures. Frequently, difficulties may be encountered in achieving disinfestation, particularly microbial contaminants that are embedded in crevices or inhabiting the interior of explants. Completion of Stage I is determined variously. With buds or terminal stem-growths as explants, the criterion is moderate enlargement to enable transfer to Stage II. Sometimes, it is desirable that adventitious organs are generated from the explant e.g., shoots from leaf segments, bulbs from bulb sections. The newly-formed organs are separated from the explant and their further culture proceeds through Stage II. In still other instances, a vigorously-growing callus signals fulfillment of Stage I. Generally, Stage I is completed in 4–6 wk.

Mother block concept: A slowly-multiplying indexed and stabilized set of cultures called as mother block. It serves as a source of cultures (explants) for Stage II multiplication.

3. Stage II: Shoot production

Steps in this stage are:

a. Subculture shoot cultures at 4–5 wk intervals.

b. Three to eight-fold increase in shoot numbers (4.3×10^7 shoots/explant/year)

c. Number of possible subcultures is dependent on the species/ cultivar

Propagule multiplication is accomplished in the second stage. The plant materials from Stage I are now subcultured repeatedly until desired quantities of the embryos, shoots or other organs are obtained. The bulk of the culture activity and time will be spent in this stage.

4. Stage III: Pretransplant (rooting)

Difficulties are usually encountered when the propagules from Stage II are transplanted directly into the soil. In Stage III, the unrooted shoots are rooted. The plants are rendered capable of autotrophic development, and adaptable to the change in moisture stress and light intensity. They are also conferred a degree of resistance to common microbial infections. Bulbs, corms, tubers, and other similar storage organs are fulfilled of their dormancy requirements. Stage III is generally accomplished in one passage and lasts for 2–4 wk. Steps involved in Stage III are:

a. Preparation of Stage II shoots/shoot clusters for transfer to the soil (prehardening)

b. Elongation of the shoots prior to *ex vitro* rooting

c. Fulfilling dormancy requirements of storage organs

5. Stage IV: Transfer to natural environment

Ultimate success of the shoot cultures depends on the ability to acclimatize vigorously-growing quality plants from *in vitro* to *ex vitro* conditions.

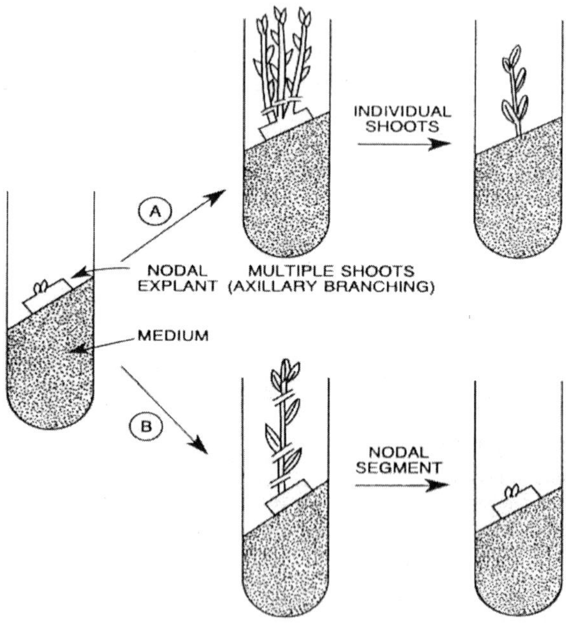

Fig. 22 Micropropagation by induction of axillary buds to develop into shoots: (A) With axillary branching and (B) Without axillary branching

Acclimatization is a process by which *in vitro* grown plantlets adjust physiologically and anatomically from *in vitro* to *ex vitro* conditions. It is relatively a slow process and may take weeks. The plantlets must adjust from high to low relative humidity (e.g., from 98–99% to 20–60%). The *in vitro* grown plantlets have poor cuticle development, and epicuticular wax needs to be formed. They have abnormal stomatal development and function (either are not properly depressed or do not open and close properly). These also have non-functional or poorly-functional roots. They must adjust from low-light to high-light intensity, from low photosynthetic competence (heterotrophic nutrition) to photosynthetic competence. The *in vitro* developed plantlets have poorly-developed leaf structure and poorly-developed chloroplasts. There may be culture medium carryover effects. Dormancy may need to be overcome. Acclimatization structures include plastic covers, humidity tent, overhead mist and fog system.

8.6. BUDS CULTURE

Buds contain active or quiescent meristems depending upon physiological state of the plant. Most vascular plants have an indeterminate mode of growth in which the leaf axils contain subsidiary meristems. They grow into a shoot that is identical to the main axis. Only a limited number of axillary meristems develop depending on the degree of branching displayed by a particular species. Different types used in bud culture are as follows:

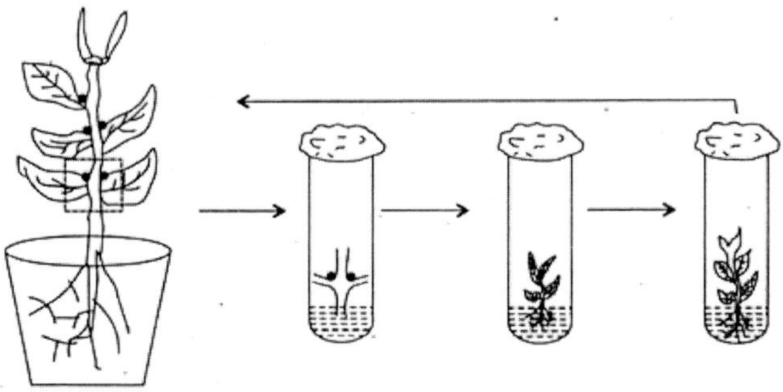

Fig. 23 Schematic representation of vegetative propagation by single node culture

8.6.1. Single Node Culture

This is the most natural method in vegetatively-propagating plants *in vitro* since it is also applicable *in vivo.* The bud found in axil of leaf is similar to the bud found in stem tip (Fig. 23). The bud is isolated and cultured on a nutrient medium for *in vitro* development. The buds found in the axil of newly-formed leaves can also be subcultured and allowed to develop into plantlet. When enough shoots have been obtained, these must be rooted and then transferred to the soil. In this technique, no cytokinin is added where buds and shoot tips are isolated and cultured. The single node method has a good success in *Vitis, Rosa, Eucalyptus grandiflora, Araucaria cunnighammii,* potato, tomato, cucumber etc.

8.6.2. Axillary Bud Method

In this method, shoot tip is isolated, from which the axillary bud in the axils of leaves develop under the influence of a relatively-high concentration; because the apical dominance is stopped by high concentration of cytokinins (Fig. 24). If the number of axillary shoots or side shoots forms from shoot tip, they can be inoculated on to a fresh medium containing high cytokinin. This is rarely used for carnation-type plants; while for strawberries and gerbera plants, it is used on a large scale.

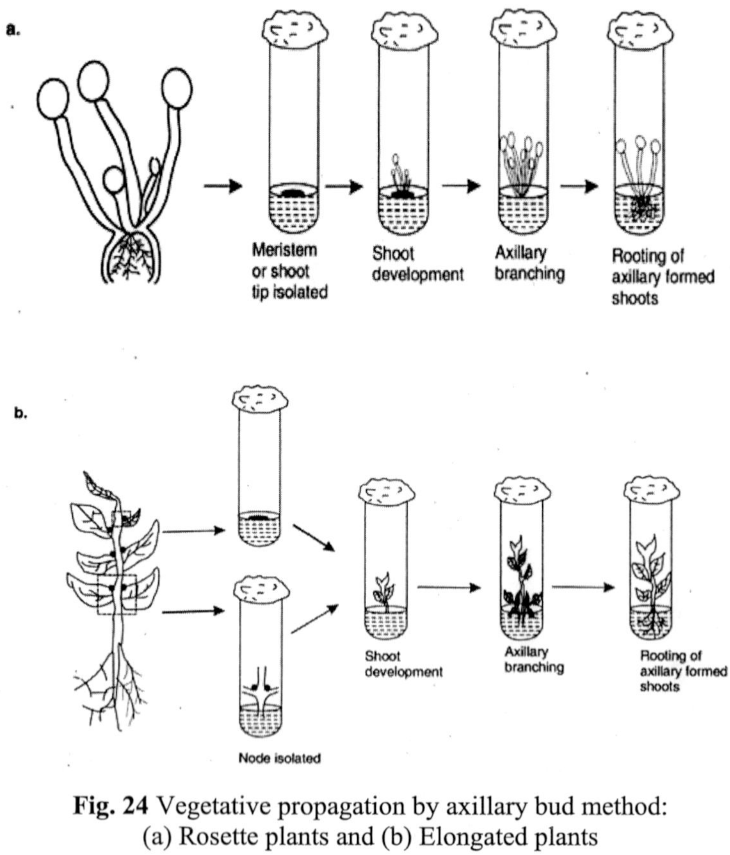

Fig. 24 Vegetative propagation by axillary bud method: (a) Rosette plants and (b) Elongated plants

8.7. FACTORS AFFECTING MICROPROPAGATION

Source of explants

Totipotentiality is probably a characteristic of all the plant cells, but its expression may be limited to particular cells. Totipotency is confined to the cells that are meristematic. The explant need not contain meristemoids, provided some of its cells are capable of generating meristemoids when suitably stimulated. When efforts with explants from all other plant parts fail, it may still be possible to generate plants from the shoot apex. Other meristems, including cambium and intercalary meristems are applicable with some genera. Familiarity with a cultivar's peculiarities is frequently helpful in seeking explant sources. Thus, sections of the leaves might be employed with plants that normally regenerate from leaves. Similarly, sections of bulb, root, stem, flower, ovary, nucellus, cotyledon and other structures may have applicability with different species.

Developmental phase of explant source

Woody perennials are also clonally propagatable through tissue cultures. However, success has required the use of explants from juvenile sources. Fortunately, juvenile growths can be found even in mature plants of many species. In trees that originate in seedlings, juvenile growths can be observed near the juncture between main stem and roots or as sprouts from roots; severing the tree at its base is often followed by flushes of juvenile shoots from the stump.

Age of explant source

Whether they are juvenile or other normally-regenerative sources, the degree of success is related inversely to their age. Among the trees, the regenerating tendency of tissue cultures diminishes with each year of plant maturation. Tissues excised from the more-recently produced parts of the herbaceous plant are also more regenerative than those from the older regions. Furthermore, embryogenesis in somatic cells has been generally associated with cultures established from embryo explants. In *Zea mays*, Greene and Phillips observed both callus and plants from young caryopsis whereas embryo explants from mature kernels produced only the callus.

Preculture treatment of explant source

With cultivars that have recurring dormancy, the climatic requirements that underlie the dormancy must be satisfied, before tissue cultures are attempted. Regeneration of the bulbs will require the use of properly-chilled bulbs as the source of explants. Photoperiod needs may also be characteristic of the explant source.

The explant size

The probability of recovering virus-free plants from infected stock is greater with smaller shoot-apex explants. However, the number of explants that survive cultures and the rates of their growth and plant multiplication *in vitro* are directly dependent on the initial size. But, the production of genetically-aberrant plants might be higher with larger shoot-tip explants.

Chemical composition

The MS medium is generally good for the propagation of many plants. The basic medium may be supplemented with an auxin to stimulate rooting. Addition of 0.1 mg/l KN and 0.3 mg/l 2,4-D renders the medium effective in callus induction. The MS salts with 170 mg/l sodium dihydrogen phosphate ($NaH_2PO_4H_2O$), 80 mg/l adenine sulfate dihydrate, 3% sucrose, and 30 mg/l N_6 r-r-dimethylallyl-aminopurine (2iP) generally stimulates axillary branching. Murashige and Skoog's salts with 2 mg/l each of IAA and KN are good for the induction of adventitious shoots.

Indeed, manipulation of organogenesis through alterations of the auxin-cytokinin balance has been achieved in cultures of many mosses, ferns, gymnosperms, dicots and monocots.

Physical qualities of the medium

The choice between agar and liquid media determines the extent of explant survival, as well as of plant multiplication and growth. Firmness of the gel is influenced by the pH of nutrient medium; a more fluid agar is attained at lower pH and harder gels are resultant at higher pH.

Liquid formulations are employed with or without agitation. Cultures in stationary liquids might be improved by providing some inert material to support the tissues. Filter paper bridges or platforms and glass wool have been generally satisfactory. Vigorous aeration may be necessary when large volumes of the liquid media are used in the stationary cultures. Agitation of the liquid cultures is achievable variously. In cloning plants through the shoot multiplication process, slow mixing rates may be preferred over vigorous churning. Shakers offer vigorous agitation, as is desired for cultures of the cells in suspension. Moderate agitation rates are typical of reciprocating shakers whereas the highest speeds are associated with gyratory shakers.

A pH near 5.0 has been used routinely with liquid formulations and 5.8 for agar-gelled media. Precipitation of some nutrients is associated with higher pH. On the other hand, poor gellation of agar results with decreasing pH. Small volumes of the media, particularly of the liquid formulations, are advisable when initiating cultures from small explants.

Light requirements

Growth and organized development of the plant tissue cultures are profoundly influenced by light. During the stages of culture initiation and propagule multiplication, an intensity of 1,000–3,000 lux has been generally considered as optimum. It is preferable that the light intensity be increased to 3,000–10,000 lux to prepare plants for their transfer to soil i.e., in Stage III. The increased intensity may retard the growth and induce slight chlorosis *in vitro*. Nevertheless, when such plants are transferred to the soil, they will survive with higher frequencies, and grow more vigorously than the greener and taller plants that are obtained at lower light intensities.

Fluorescent lamps with high emittance in the blue and red regions are probably desirable. Blue light is the most effective in stimulating shoot differentiation. In contrast, rooting is enhanced markedly by red light. Photoperiodism, in strict sense, may not be involved, but organogenesis *in vitro* is influenced by diurnal light/dark cycles. A photoperiod of 16 h has been generally satisfactory for shoot and root initiation in a variety of plants. The formation of tubers and other storage organs may require longer dark periods, since they are normally short-day processes.

Temperature requirements

Warm temperatures, about 27°C, have been optimum for tropical genera while cool temperatures, near 20°C, have been found good to alpine plants. Many bulb cultivars have chilling requirements, prior to initiation of their tissue cultures. Bulbs developed *in vitro* must also be chilled to ensure their successful transfer to soil.

Atmospheric components

Premature deterioration of the tissue culture is sometimes attributed to the diverse gases released by the cultures. Charcoal filters can be used to exclude smog constituents in highly-urban surroundings. Filters are necessary when the ambient air contains high concentrations of dust and spores. Relative humidity is a concern for extreme climates. The high moisture content of the tropical air often favours the development of fungi and bacteria, and reduction of the relative humidity with dehumidifiers is desirable. Rapid drying of the media may be a problem in arid climates. This is easily resolved by employing more tightly-fitting culture-vessel closure.

8.8. MICROPROPAGATION OF ORNAMENTAL PLANTS

Saintpaulia, Begonias, *Streptocarpus*, *Gloxinia* and members of the family *Gesneriaceae* are propagated by adventitious buds, and they exhibit high rates of bud induction. Culture requirements are generally not stringent, but the medium needs to be reinforced by a cytokinin/auxin.

By and large many Bromeliads are propagated by shoot-tip culture, and by adventitious and axillary buds in several instances (*Billbergia, Guzmania, Tillandsia, Uriesia, Ananas* etc.). Regeneration via callus and callus cell-suspensions are also feasible, though uniformity of the progeny is not guaranteed. Ploidy plantlets may arise showing phenotypic variability, which may sometimes form the basis for better selection of the plants showing novel traits.

Several bulb plants (bulbous monocotyledons) develop adventitious buds and bulbils from pieces of scale leaves or basal tissue *in vitro*. *Allium*, *Narcissus*, *Gladiolus*, *Nerme, Freesia, Tulipa, Ornithogalum* and *Cyclamen* exhibit high propensity for regeneration by proliferation of axillary shoots from leaf bases. Rapid propagation of *Zephyranthes* (Philomena *et al.,* 1987) and of allied species of *Lilium, Amaryllis, Iris* etc. through tissue culture has been possible. Hussey (1982) recommends periodic trimming of developing bulbils, a practice that would encourage development of waves of bulbils. Plant regeneration from *Trillium* spp. *in vitro* has been successful (Pence and Soukup,

1986). A technique has been described for micropropagation of spray carnations using axillary buds (Ghosh and Mohan Ram, 1986).

Anthuriums are in great demand and cloning of the plants through tissue culture is widely practised in Netherlands (Pierik, 1970; van der Meiss, 1980). Propagation is mostly by adventitious buds (Leifring and Soede, 1978, 1979). Being tardy in growth and prone to contamination, large-scale multiplication as a commercial proposition has not, however, been encouraging.

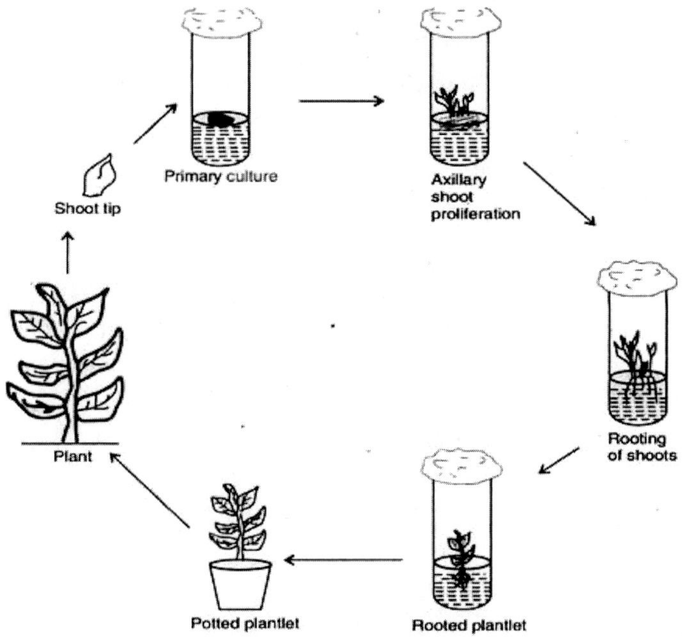

Fig. 25 Schematic representation of shoot-tip culture

8.9. ADVANTAGES OF MICROPROPAGATION

In vitro propagation techniques are preferred to conventional practices of asexual propagation because of the following advantages with regard to crop improvement:

- A small amount of tissue is needed as explants for regeneration of millions of clonal plants in one year. It would take years to propagate an equal number of plants by conventional methods.

- Micropropagation technique provides alternative method, which helps in bulking up new cultivars of important trees rapidly that would otherwise take years by the conventional methods.
- The *in vitro* stocks provide year-round nursery for ornamental, fruit and tree species, and can be quickly proliferated at any time of the year.
- The sterile status of the cultures eliminates the danger of disease introduction by reducing the period of quarantine and speedy international exchange of the plant material.
- Meristem tip culture provides disease-free plants in *Dianthus, Chrysanthemum, Gladiolus* and *Pelargonium.*
- **Seed production:** Micropropagation can be used for seed production in crops where a limiting factor is the high degree of germplasm conservation required. For example, the production of male sterile lines in onion to provide an alternative to difficult back-crossing methods, and the production of *Asparagus* for producing high-quality supermale and female homozygous lines from which all the desirable male hybrids can be produced.
- **Germplasm storage:** Micropropagation assures the availability of bioresources as the need arises. Meristem can be expected to survive the cryopreservation procedures and it is an excellent material for germplasm preservation of crop species.

■■■■

9 HAPLOID PRODUCTION

9.1. INTRODUCTION

The term 'haploid' refers to those plants, which possess a gametophytic number of chromosomes (single set) in their sporophytes. The interest in haploids largely stems from their considerable potential in plant breeding, especially for the production of homozygous plants and for the detection of mutants.

The significance of haploids in the fields of genetics and plant breeding has been realized for a long time. However, their exploitation remains restricted because of low frequency (0.001–0.01%) with which they occur in nature. Spontaneous production of haploids usually occurs through the process of apomixis or parthenogenesis (embryo development from an unfertilized egg). Artificial production of haploids was attempted through distant hybridization, delayed pollination, application of irradiated pollen, hormone treatment and temperature shock. However, none of these methods was dependable and repeatable. It was in 1964 that Guha and Maheshwari reported the direct development of embryos and plantlets from microspores of *Datura innoxia* by the culture of excised anthers. Later, Bourgin and Nitsch (1967) obtained complete haploid plants of *Nicotiana tabacum*. Since then, anthers containing immature pollen have been successfully cultured for a wide range of economic important species.

9.2. ANTHER CULTURE

As said earlier, Sipra Guha and Satish Chandra Maheshwari in 1964 cultured the anthers of *Datura innoxia* with the objective of understanding the physiology of meiosis. However, they made a chance observation of development of embryo-like structures in these cultures.

Later on, they (Guha and Maheshwari, 1966, 1967) confirmed that these embryos were haploid and derived from pollen grains. Due to the usefulness of the haploids, this report has revolutionized the whole plant breeding programmes. Development of the embryos from male cells i.e., pollen grains, is known as androgenesis.

9.2.1. Technique of Androgenesis

Guha and Maheshwari (1964, 1966) cultured anthers of *Datura innoxia* on a BM supplemented with CM or KN.

Young flower buds at the appropriate stage of pollen development are surface-sterilized and the anthers are removed along with their filaments. The anthers, after a quick dip in absolute ethyl alcohol, are excised under aseptic conditions and placed horizontally on the medium. Sufficient care should be taken that the anthers are not injured during the process of anther culture, otherwise wounding causes callusing of anther wall tissues. Nutrient medium solidified with agar has been used (Guha and Maheshwari, 1964) for most species. However, in species of *Nicotiana* and *Hyoscyamus*, a better yield of pollen-plantlets is obtained by using liquid medium. When dissection of the buds poses problems, intact whorl of the stamens as in tobacco or whole spike as in barley may be cultured.

9.2.2. Development of Androgenic Haploids

Androgenic haploids may develop in four ways (Bhojwani and Bhatnagar, 1990) (Fig. 26).

1. **Pathway I:** Microspore divides by equal division forming two identical daughter cells. Both these cells contribute to the development of sporophyte (e.g., *Datura innoxia*). In this pathway, vegetative and generative cells are not formed.
2. **Pathway II:** Microspore undergoes unequal division resulting in a large vegetative cell and a small generative cell. The sporophyte develops from the vegetative cell alone (e.g., *Nicotiana tabacum, Hordeum vulgare, Capsicum annum, Triticale*). The generative cell may or may not divide further, but degenerates eventually.
3. **Pathway III:** Microspore undergoes unequal division resulting in a large vegetative cell and a small generative cell. The

sporophyte develops from the generative cell alone (e.g., *Hyoscyamus niger*).

4. **Pathway IV:** Microspore undergoes unequal division forming a large vegetative cell and a small generative cell. Both the vegetative and generative cells contribute towards the formation of sporophyte (e.g., *Datura metel*, *Atropa belladonna*) (Fig. 24).

Irrespective of the nature of first pollen mitosis, further development may lead to the formation of embryoids and plantlets or a mass of callus (Fig. 27). In *Nicotiana*, *Datura*, *Atropa*, *Capsicum*, *Petunia*, *Brassica* and *Solanum,* the pollen grains produce embryoids and plantlets directly without the intervention of a callus phase. In many of the monocotyledonous plants such as *Oryza*, *Triticum*, *Hordeu*m and *Lolium,* no embryoid formation ever occurred; instead, the pollen grain developed into an unorganized mass of callus, which could be manipulated by transfer to a suitable regenerating medium to differentiate into plantlets.

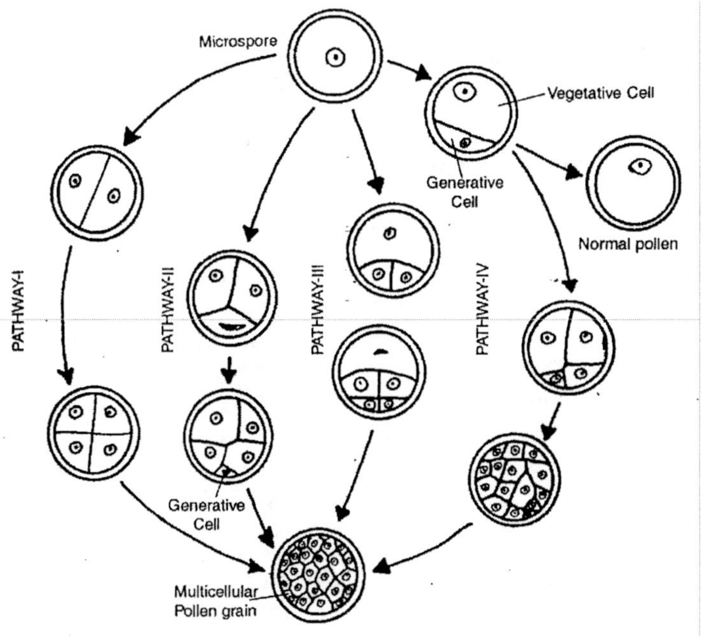

Fig. 26 Diagram showing development of multicellular pollen grains (Bhojwani and Razdan, 1983)

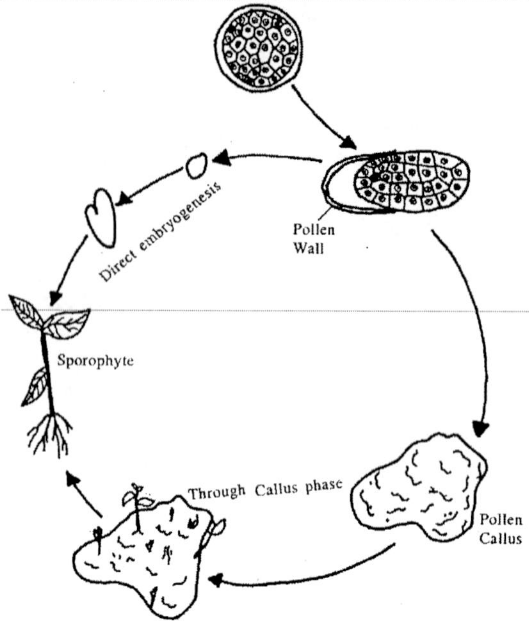

Fig. 27 Diagram showing origin of sporophyte from pollen grain (Bhojwani and Razdan, 1983)

9.2.3. Factors Affecting Anther Culture

The factors that affect androgenesis are as follows:

- **Genotype of the donor plant:** The genotype of the donor plant plays a significant role in determining the frequency of androgenesis. In wheat, the induction of pollen callus and green plantlet formation is controlled by many genes. In *Hordeum,* each genotype differs with respect to androgenic response in anther culture. Hence, the genotype that responds well in anther culture should be selected. It has been possible to improve the androgenic response of poorly-responding tobacco and barley by following breeding approach.

- **Anther wall factor(s):** The effect of anther tissues on the ontogeny of embryoids is not fully understood. Histological studies support the role of anther wall factor(s) in pollen embryo development. In henbane, the embryogenic pollens are restricted to the periphery of the anther locule suggesting that a gradient of critical substances of tapetal origin is instrumental in induction of embryogenic division in the pollen grains. Glutamine, L-serine and myoinositol could replace wall factor(s) for isolated pollen cultures.

- **Culture medium:** Culture medium plays an important role in inducing divisions in microspores. Basal media of Murashige and Skoog (1962), and Gamborg *et al.* (1968) are especially advantageous for many species. However, N_6 medium is used for cereal anther culture. Although usually 2% sucrose is added to the medium, anthers of barley and *Brassica* require a higher concentration (10%) of sucrose. Similarly, in anther cultures of wheat 6% sucrose promoted pollen callusing. However, continuous culture in a high-sucrose medium proved deleterious to the normal differentiation of embryoids into plantlets unless the embryoids were transferred to a low-sucrose medium.
  - Iron in the medium is reported to be indispensable for the production of plantlets from microspores. AC in the medium has been reported to have stimulatory effect on the embryoid formation and plantlet regeneration in tobacco. Charcoal possibly absorbs the inhibitory substances present in the anther wall.
  - Most solanaceous species exhibit androgenesis on a medium containing mineral salts, vitamins and sucrose. In majority of the non-solanaceous plant species that are known to exhibit androgenesis, formation of pollen-embryo occurs only when the medium is fortified with growth adjuvants such as yeast extract, CH, CM etc. For some others, addition of auxin (2,4-D, IAA, NAA) or cytokinin (KN) promoted androgenesis.
- **Developmental stage of anther:** The developmental stage of microspores at the time of culture is a critical factor for the induction of division. Anthers of some species give the best response if the pollen is cultured at first mitosis or later stages (post-mitotic - *Datura innoxia, Nicotiana tabacum, Petunia hybrida*) whereas in most others, anthers are more productive when cultured at uninucleate microspore stage (pre-mitotic stage). The early bicellular stage is best for *Nicotiana sylvestris* and *Atropa belladonna*, but is absolutely essential for *Nicotiana knightiana*. Young anthers showing meiotic stages in pollen

mother cells and older ones with starch-filled grains failed to produce embryoids. However, there are certain exceptions to this rule. The ideal stage for androgenesis in *Arabidopsis thaliana* and *Lycopersicon esculentum* is when the microspore mother cells are in meiosis-I whereas *Brassica* species show better response when mature anthers or pollen isolated from them are cultured. External markers, such as the length of the stamen, can be used for knowing the stage of anther.

- **Physiology of the donor plant:** Age of the plant influences the androgenic response in anther culture. The buds developed towards the beginning and peak of the flowering season respond better than those produced later in the season. Anthers taken from the plants grown under short-day (8 hr/d) conditions show relatively better response than long-day (16 hr/d) donor plants. Growth conditions of the donor plants also affect the embryogenic potential of the cultured anthers. An increase in anther response and embryo yield could be obtained in tobacco by subjecting the anther of donor plants to nitrogen starvation.
- **Effect of temperature and light:** A cold treatment given to the flower buds before excising the anthers for culture enhances the induction frequency of microspore androgenesis. In tobacco and *Datura,* the buds treated at 5°C for 72 h induced 58% of anthers to yield pollen embryoids while buds treated at 22°C for the same period, the response was only 21%. In rye, maize and *Pennisetum*, the induction of androgenesis has been found very effective if they are stored at low temperatures prior to anther culture.
- In contrast to this, high temperature treatments increased the response of cultured anthers in *Brassica napus* whereas in *B. juncea*, a combination of cold pretreatment and high-temperature shock was essential for inducing embryogenesis.

9.2.4. Procedure for Anther Culture

The culturing of anthers containing immature microspores is called anther culture and is as follows (Fig. 28):

1. Collect unopened flower buds from the plants.
2. Surface-sterilize the flower buds with tween 80 and mercuric chloride solution. Then wash them with distilled water.

3. Carefully take the anthers from the flower buds and keep them in a petri dish.
4. Select the anthers, which have just completed first meiotic division, by acetocarmine squash test.
5. Aseptically transfer suitable anthers to a liquid or solid nutrient medium containing glutamine, L-serine and inositol.
6. Incubate the culture plates at 25°C for about 15 d and the anthers grow into embryoids.
7. Transfer the embryoids to a rooting medium in the culture plates and incubate at high lux light. The embryoids become plantlets within 4–5 wk.
8. Plant the well-developed haploid plantlets in the pots in a greenhouse for acclimatization.

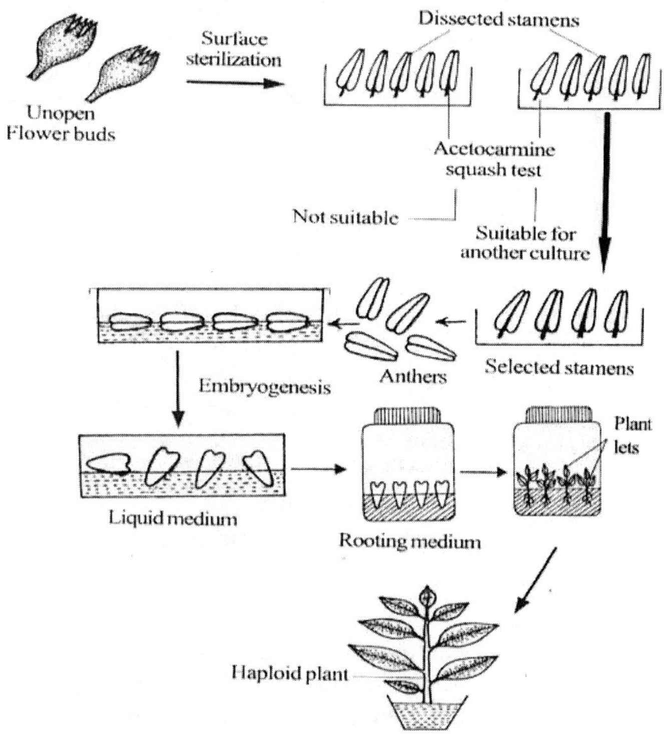

Fig. 28 Stages of anther culture

9.3. POLLEN CULTURE

Anther culture has some disadvantages. The plants obtained from anther culture may originate not only from pollen grain, but also from various parts of the anther, resulting in mixed ploidy. Anther wall may produce inhibitory substances, which are not conducive to pollen embryogenesis. The pollen population within an anther is genetically heterogenous. Therefore, the plants arising from an anther constitute a heterogenous population. Moreover, where the formation of haploid plants is preceded by callusing of pollen, the tissue arising from an anther is actually a chimera because numerous pollens in the anthers contribute to the callus. These problems can be overcome by isolated pollen culture. In addition, the problem of competition between a large numbers of pollen grains for growth in the limited available space within the anther can also be solved through isolated pollen culture.

Pollen grains when isolated and grown *in vitro,* give rise to haploids and callus, and this approach is called as pollen culture. The first report on the tissue formation from isolated pollen of an angiosperm was from Kameya and Hinata (1970). They used hanging drops to culture the mature pollen isolated from *Brassica oleracea* and the hybrid *B. alboglabra* X *B. oleracea* at low temperature. Pollen may be isolated either by squeezing or float culture of the anthers (Fig. 29).

- **Squeeze method:** Nitsch (1974) described the method of raising haploid plants from isolated pollen grains of tobacco. Unopened flower buds are surface-sterilized. About 50 anthers are placed in a beaker containing about 20 ml of the liquid medium. Pollen grains are squeezed out by pressing the anther with the help of a glass rod or syringe piston. The whole solution is filtered through a nylon sieve. The suspension, thus obtained, is centrifuged at 500–800 rpm for 5 min. The pellet containing pollen is resuspended in a fresh medium and the process is repeated twice. Finally, the pollen is mixed with a suitable culture medium and the suspension is pipetted into petri plates. The cultures are stored at around 25°C in diffuse light (500 lux).

- **Nurse-culture method:** Nurse-culture technique was found to be suitable for raising haploid tissue clones from isolated pollen grains. Intact anthers are obtained from a flower bud and placed horizontally on the surface of a semi-solid medium. A filter paper disc is placed over the anther. A suspension of pollen from anthers of another bud was separately prepared in a liquid medium and about 10 pollen grains are transferred to each filter paper disc. Within a month, numerous clones were formed.

Callus derived from the floral parts replaces the nurse tissue. The extract of cultured anthers has equal nursing effects on the growth and development of microspores. A purely synthetic medium developed for raising haploid plants from isolated pollen grains of tobacco and *Datura* may also be suitable. The crucial ingredients of this medium are glutamine, L-serine and inositol. Haploids from isolated pollen culture have been reported in tomato, maize, *Petunia hybrida* and maple.

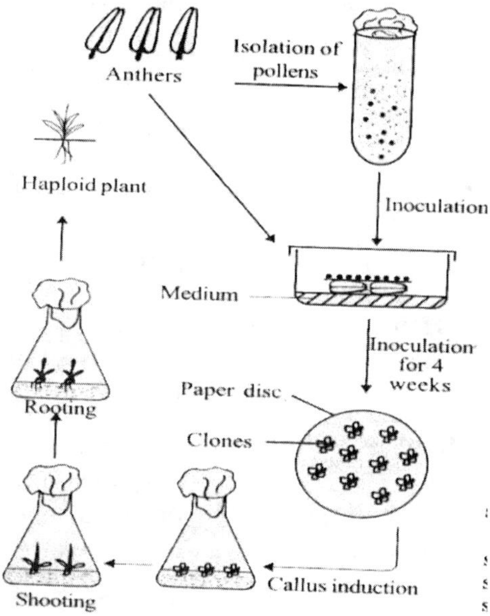

Fig. 29 Pollen culture using nurse-culture technique

9.4. COMPARISON BETWEEN ANTHER AND POLLEN CULTURES

Anther culture is easy, quick and practicable. Anther walls act as conditioning factors and promote culture growth. Thus, anther cultures are reasonably efficient for haploid production. The major limitation is that the plants not only originate from pollen, but also from other parts of anther. This results in the population of the plants at different ploidy levels (diploids, aneuploids). The disadvantages associated with anther culture can be overcome by pollen culture.

Many workers prefer pollen culture, even though the degree of success is low, as it offers the following advantages:

- Undesirable effects of anther wall and associated tissues can be avoided.
- Androgenesis, starting from a single cell, can be better regulated.
- Isolated microspores (pollen) are ideal for various genetic manipulations (transformation, mutagenesis).
- The yield of haploid plants is relatively higher.

9.5. SIGNIFICANCE OF USES OF HAPLOIDS

- **Development of pure homozygous lines:** Haploids are most useful as a source of homozygous lines. By the use of anther/microspore culture, the period can be reduced to few months to year that would take 6–8 y in a conventional breeding. Homozygosity is important for those plants, which have long juvenile phase (period from seed to flowering) such as fruit trees, forest trees and bulbous plants.
- **Hybrid development:** As a result of complete homozygosity obtained from diploidization of haploids, one can rapidly fix traits in the homozygous condition. These lines can be used for the production of pure F_1 hybrids.
- **Induction of mutations:** Haploid cell cultures are useful material for studying somatic cell genetics, especially mutation and cell modification. Mutants, which are resistant to antibiotics, herbicides, toxins etc. have been isolated in a number of plant species. Carlson regenerated mutant plants of *Nicotiana tabacum* that showed resistance to *Pseudumonas tabaci.*
- It is possible to obtain only male plants, which have higher productivity and yield than female plants by chromosome doubling. For example, *Asparagas officinalis* in which androgenic haploids produced from male asparagus plants are

either X or Y. Chromosome doubling of Y results in supermales (YY), which can be vegetatively propagated. If XX plants are crossed with YY plants, the progeny will be all male. Using this method, Thevenin and Dore (1976) established high-yielding male lines of *Asparagus*.

- Haploids have been used in the production of aneuploids, monosomics in wheat, trisomics in potato and nullisomics in tobacco.
- By anther culture, not only the haploids, but also the plants of various ploidy levels and mutants are obtained, and can be incorporated into the breeding programme.
- The haploid method of breeding involving hybrid sorting from F_1 hybrids through anther culture is considered superior over pedigree and bulk methods. For example, the frequency of homozygous F_1 gametes is one in one hundred, and then the frequency of homozygous F_2 plants derived from the fusion of two such superior similar gametes would be one in ten thousand. Therefore, smaller population of double haploids derived from haploids will need to be handled.
- Haploid production has been used for the introduction of disease-resistant genes into cultivars. F_1 or F_2 anthers of one cultivar developed with disease resistance are plated and haploids are developed. These haploids are screened for resistance and then diploidized. Foroughi and Friedt (1984) developed barley variety resistance to yellow mosaic virus. Witherspoon *et al.* (1991) reported a new variety of tobacco, which shows resistance to highly-necrotic strain of potato virus Y. Other salt-tolerant, insect-resistant varieties are also possible to develop through haploids.
    - Recent application of DH lines is their use in genome mapping for major genes and/or quantitative traits in barley, rice, oilseed rape etc. In a population of DH lines, the identification of markers is much more secure, as most intermediate phenotypic expressions are excluded due to heterozygosity.
- **Developing asexual lines of trees and perennial species:** Chinese workers obtained a pollen-derived rubber tree taller by 6 m, which could then be multiplied by asexual propagation to

raise several clones. Similarly, in haploids of poplar, desired genotypes are selected and then doubled after 7–8 y. This coincides with their flowering, enabling them to be propagated both asexually and sexually. Pollen-derived haploid plantlets have also been obtained in *Citrus microcarpa*, *Vitis vinifera, Litchi chinensis*, *Camellia sinensis* etc.

- Pollen embryos, being haploid, can be used as tools for genetic transformation studies. Neuhaus *et al.* (1987) produced transgenic plants of *Brassica napus* by microprojectile bombardment of DNA into individual pollen embryos. Haploid plants regenerated from transformed embryos on diploidization expressed the recombinant transgene homozygously. The PEG-induced uptake of foreign DNA by isolated microspores of *Hordeum vulgare* also showed potential for *in vitro* androgenesis comparable to untreated controls. These studies suggest the possibility to produce stable transformants using pollen embryos or microspores.
- *In vitro* production of haploids can solve some problems in genetic studies since gene actions are readily manifested due to a single allelic dose present in the chromosomes of an entire genome.

9.6. LIMITATIONS

- High level management and expertise is required to operate the tissue culture production of haploids.
- Selective cell division must take place in the haploid microspores and not in other unwanted diploid tissues. This selective cell division is often impossible.
- The high incidence of albinism in some types of anther and pollen culture
- There will be lack of selection of traits during the derivation of haploid material.
- The doubling of a haploid does not always result in the production of a homozygote.

9.7. GYNOGENESIS

Development of the plants from unfertilized cells of the female gametophyte (embryo sac) in floret, ovary or ovule culture is becoming an attractive alternative to anther culture for haploid production. Production of haploid individual by the development of an unfertilized egg cell is known as gynogenesis. This can be achieved through:

- Delayed pollination
- Through the use of abortive pollen
- Pre-exposed to ionizing radiation
- By using alien pollen

Efforts to raise parthenogenetic haploid plants using *in vivo* techniques yielded little success. San Noeum in 1976 was the first to demonstrate that gynogenesis, an essentially *in vivo* phenomenon, can be induced under *in vitro* conditions. Gynogenic haploids have been obtained in *Hordeum vulgare* using ovary culture (San Noeum, 1976). Zhu and Wu (1979) obtained haploid plants from cultured, unpollinated ovaries of *Triticum aestivum* and *Nicotiana tabacum*.

Young flowers, ovaries or ovules have been used as the explant to produce gynogenic haploids. Generally, ovules attached to the placenta respond better than isolated ovules. In *Gerbera jamesonii* so far, only isolated ovule culture has been successful. In *Helianthus annuus,* ovule culture proved superior to ovary or floret culture (Yan *et al.,* 1988). In *Beta vulgaris,* ovule culture has been mostly used although Van Geyt *et al.* (1987) have reported that ovary culture was better. In rice, the best results of gynogenesis were obtained when unhusked flowers with pistil and stamens attached to the receptacle were inoculated on a liquid medium. The response was less effective when stamens were removed and was worst when single pistils were used. Similarly, in barley, whole florets (with or without stamens) inoculated vertically on a solid medium performed better than randomly-placed single pistils (Huang *et al.,* 1982).

The stage of the female gametophyte at culture is crucial. Mostly, the explants cultured at nearly-mature embryo-sac stage gave best results (Yang and Zhou, 1990; Yang *et al.,* 1986). Rice is an exception, where inoculation of ovaries at 1–4 nucleate stage of female gametophyte development proved most responsive (Zhou *et al.,* 1986).

A beneficial role of cold treatment on gynogenesis has been reported. Pretreating the capitula of sunflower at 4°C for 24–48 h before culture significantly increased the induction frequency (Yan *et al.*, 1987). Cai *et al.* (1988) observed a promontory effect of cold treatment of the young panicles of rice at 7°C for 1 d before ovary culture. In *Cucumis melo,* pollination of the pistils with irradiated pollen was essential to obtain ovules capable of forming gynogenic haploids (Katoh *et al.*, 1993).

The culture media used for the production of gynogenic haploids vary considerably. The most widely-used BM in these studies happens to be N_6. However, N_6 and MS media were equally good for the production of maternal haploids of maize. Generally, sucrose is used at higher levels: 3–10% for barley, 8–14% for wheat, 5–12% for maize and 3–6% for rice and gerbera.

The induction medium for the graminaceous species is generally supplemented with 2,4-D (2 mg/l) or MCPA (0.125–0.5 mg/l). The best combination of growth regulators for parthenogenic haploid production in ovary/ovule culture of *Beta vulgaris* was found to be 0.3 mg/l BAP. On the other hand, a hormone-free medium gave higher induction frequency than with MCPA.

Mostly, solid media have been used to induce gynogenesis. However, in rice, at the induction stage, liquid medium proved superior to solid medium. On the other hand, for maize, solid medium was more productive than the liquid medium in the dark.

Gynogenic haploids have also been obtained from cultured unpollinated ovules in sunflower, potato, *Allium cepa, Beta vulgaris, Brassica oleracea, Cucumis melo, Gerbera jamesonii, Hevea brasiliensis, Melandrium album, Nicotiana rustica, Allium tuberosum* and *Petunia axillaris,* and unpollinated ovaries of rice, maize, *Allium tuberosum, Coix lacryma-jobi, Cucurbita pepo, Helianthus annuus, Hordeum vulgare, Hyoscyamus muticus, Lilium davidii, Nicotiana tabacum, Triticum aestivum, T. durum* and *Zea mays* (see Table 9). Flower buds have been cultured to obtain gynogenic haploids in *Allium cepa* and *A. porrum*.

9.7.1. Steps involved in the Technique of Gynogenesis

The following steps are necessary for the production of plantlets through the process of gynogenesis (Fig. 30):

1. Culture pollinated and unpollinated ovaries.
2. For haploid production, excise the flower buds 24–48 h prior to anthesis for unpollinated ovaries.
3. Remove the calyx, corolla and stamens, and surface-sterilize the ovaries.
4. Cut off the distal part of the pedicel and implant the ovary with the cut-end inserted in the nutrient medium.
5. Use the normal Nitsch's (1951) or White's (1958) or MS or N_6 inorganic salt media supplemented with growth substances.
6. Sucrose as a carbon source is essential. Maltose and lactose also have been equally favourable.
7. When liquid medium is to be employed, place the ovaries on a filter paper raft or float with the pedicel inserted through filter paper and dip into the medium.
8. Some species require simple BM and others need addition of auxin to bring about greater stimulation of growth.

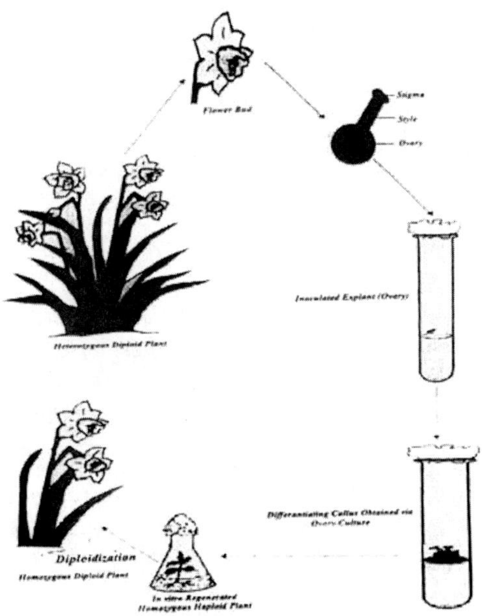

Fig. 30 Schematic representation of gynogenesis in plants

About 0.2–6% of the cultured ovaries show gynogenesis and one or two, rarely up to 8, plantlets originate from each ovary. The rate of success varies considerably with:

- **Genotype:** Markedly influenced by the genotype so that some cultivars do not respond at all. For example, in rice, japonica genotypes are far more responsive than indica genotypes.
- **Stage of ovary development:** In most cases, the optimum stage for ovary culture is the nearly-mature embryo sac. However, in rice, ovaries at free-nuclear embryo-sac stage are the most responsive. Generally, culture of whole flowers, ovary and ovules attached to placenta respond better, but in *Gerbera* and sunflower, isolated ovules show better response. Cold pretreatment (24–48 h at 4°C in rice) of the inflorescence before ovary culture enhances gynogenesis.
- **Growth regulators:** Growth regulators are crucial in gynogenesis, and at higher levels, they may induce callusing of somatic tissues and even suppress gynogenesis. Growth regulator requirement seems to depend on species. For example, in sunflower, growth regulator free medium is the best while MCPA induces somatic calli and SEs even at low level. However, in rice, 0.125–0.5 mg/l MCPA is optimum for gynogenesis.
- **Embryo sac stage:** It has been reported that complete maturation of an ovule is not necessary for induction of gynogenesis, but an appropriate stage is more important.
- **Other factors:** Sucrose level also appears to be critical. In sunflower, 12% sucrose leads to gynogenic embryo production while somatic calli and somatic embryos were also produced at lower levels. Ovaries/ovules are generally cultured in light, but at least in some species e.g., sunflower and rice, dark incubation favours gynogenesis and minimizes somatic callusing. In rice, light may lead to degeneration of gynogenic proembryos.
- **Developmental stages:** Generally, gynogenesis has two or many stages and each stage has distinct requirements. In rice, two stages *viz.*, induction and regeneration are recognized. During induction, ovaries are floated on a liquid medium having low auxin and kept in dark while they are transferred

on to an agar medium with higher-auxin concentration and incubated in light for regeneration. Haploid plants, generally, originate from egg cell in most of the species (*in vitro* parthenogenesis), but in some species e.g., rice, they arise mainly from synergids; in *Allium tuberosum* even antipodals produce haploid plants (*in vitro* apogamy). As in anther culture, gynogenesis may occur either via embryogenesis or through plantlet regeneration from callus. In rice, MCPA generally leads to a small amount of protocorm-like callus formation from which shoots and roots regenerate while picloram promotes embryo regeneration. In contrast, sugar beet usually shows embryo development while embryos regenerate following a callus phase in sunflower. In general, regeneration from a callus phase appears, at least for the present, to be easier than direct embryogenesis.

9.7.2. Advantages

- Gynogenetic haploids may be a valuable substitute for the production of homozygous lines in cases where CMS prevents the use of microspores.
- Reduction in the frequency of albino plants is observed in some species, especially cereals. In these crops, ovary/ovule culture may provide relatively-higher proportion of green haploids. For example, San Noeum and Ahmadi (1982) reported that in Barley cv. Bernice, 100% gynogenic plants were green as against 1% androgenic plants. A common problem with anther culture of *Petunia* has been high percentage of non-haploids. In contrast to this, 93% gynogenic plants of *P. axillaris* were haploid.
- It is the most efficient and reliable technique for the production of haploid and doubled-haploid plants e.g., sugar beet.
- In plants, such as sugar beet, onion, and melon, where anther culture has not been successful, gynogenesis has yielded haploids.
- In male sterile plants, the value of haploid production via gynogenesis is obvious.

9.7.3. Limitations

- So far it has been successful only in less than two dozen species.
- The frequency of responding ovaries (1–5%) and the number of plantlets/ovary (1–2) is quite low.
- Therefore, anther culture is preferred over ovary culture. Only in those cases where anther culture fails e.g., sugar beet and for male sterile lines, ovary culture assumes significance.

Table 9: List of species for which gynogenic haploids have been raised through ovary/ovule culture

Family	Species	Explant	Reference
Chenopodiaceae	*Beta vulgaris*	Ovule	Hosemans and Bossoutrot (1983); Bornman (1985); Lux *et al.* (1990)
Asteraceae	*Gerbera jamesonii*	Ovule	Sitban (1981); Cappadocia *et al.* (1988); Tosca *et al.* (1990)
	Helianthus annuus	Ovary	Cai and Zhou (1984)
Brassicaceae	*Brassica oleracea*	Ovule	Dore (1989)
Cucurbitaceae	*Cucumis melo*	Ovule	Katoh *et al.* (1993)
Euphorbiaceae	*Hevea brasiliensis*	Ovule	Chen *et al.* (1990)
Poaceae	*Coix lacryma-jobi*	Ovary	Li and Zhang (1984)
	Hordeum vulgare	Ovary	San Noeum (1976); Gu and Zheng (1984)
	Oryza sativa	Ovary	Asselin de Beauville (1980)
	Triticum aestivum	Ovary	Zhu and Wu (1979)
	Zea mays	Ovary	Ao *et al.* (1982)
Liliaceae	*Allium cepa*	Ovule	Campion and Allioni (1990); Keller (1990)
	A. tuberosum	Ovary	Tian and Yang (1989)
	Lilium davidii	Ovary	Gu and Zheng (1983); Prakash and Giles (1986)
Moraceae	*Morus alba*	Ovary	Lakshmi Sita and Ravindran (1991)
Salicaceae	*Populus simonigra*	Ovary	Wu and Xu (1984)

Scrophulariaceae	*Mimulus luteus*	Ovary	Hess and Wagner (1974)
Solanaceae	*Nicotiana tabacum*	Ovary	Zhu and Wu (1979); Wu and Chen (1982)
	Petunia axillaris	Ovule	De Verna and Collins (1984)
	Solanum tuberosum	Ovule	Tao *et al.* (1985)

9.8 BULBOSUM TECHNIQUE

Kasha and Kao (1970) discovered a new method for obtaining haploids. In interspecific crosses between tetraploids of *Hordeum vulgare* and *H. bulbosum,* nearly all the plants formed were dihaploids. Similarly, in the crosses of the diploids of these species, the progeny comprised haploids. The progeny was similar to *H. vulgare* in morphology and chromosome complement. Cytological examination has revealed that in this cross, normally, double fertilization occurs. However, selective elimination of *bulbosum* chromosomes during early stages of development of embryo, which initially contains 1 *vulgare* and 1 *bulbosum* genome, leads to haploid embryo formation (Subrahmanyam and Kasha, 1973). In this cross, the embryo, normally aborts after about 10 d of pollination, but it is possible to raise the whole haploid plant through embryo culture. Barclay (1975) has demonstrated that high-frequency haploids of wheat (*Triticum aestivum*) can be produced by crossing it with *Hordeum bulbosum*, followed by embryo culture. The genetic basis of this selective elimination of *bulbosum* chromosomes is now being investigated.

■■■■

10 NUCELLUS CULTURE, OVULE CULTURE, OVARY CULTURE AND FLOWER CULTURE

10.1. NUCELLUS CULTURE

In some plants such as *Citrus* and mango, embryos arise adventitiously from the cells of the nucellus. The nucellar embryos are of considerable importance to horticulturists as they are genetically-uniform and reproduce the maternal genotype. Another advantage of the nucellar embryos is that in contrast to clones established from the cuttings or by conventional methods of propagation, the plants derived from nucellar embryos are free from pathogens. Further, many of the desirable plant vigour and fruiting characteristics associated with juvenility are restored in the trees that are established from nucellar seedlings.

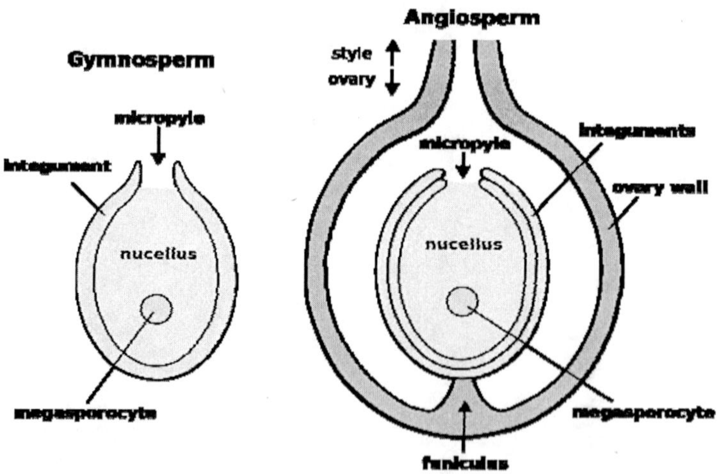

Fig. 31 Ovule and ovary of gymnosperm and angiosperm

Rangaswamy (1961) excised nucellus from pollinated ovules of *Citrus microcarpa* (a polyembryonate species) and cultured it on modified WM supplemented with CH. The nucellus proliferated into a callus and it differentiated into numerous tumeroid outgrowths called pseudobulbils, which eventually developed into plantlets. Sabharwal (1963) reported similar results in *Citrus reticulata*, but opined that the pseudobulbils develop from nucellar embryos.

Rangan *et al.* (1968, 1969) has shown that it is possible to induce nucellar embryogenesis in monoembryonate *Citrus* species (*C. grandis*, *C. limon*, *C. reticulata* and *C. sinensis*). In this investigation, the nucellus was free from zygotic embryo, and cultured on a medium supplemented with malt extract.

Pollination and fertilization are generally considered as essential prerequisites for the induction of nucellar embryogenesis. However, Button and Bornman (1971) have demonstrated that nucellus from unfertilized ovules can also form embryoids. The nucelli from the pre-fertilized ovules exhibit embryogenesis only if the medium is supplemented with malt extract and adenine.

Mitra and Chaturvedi (1972) induced embryogenesis in nucelli obtained from unfertilized ovules of *Citrus sinensis* and *C. aurantifolia*. They reported that the embryos developed directly or from callus, which proliferated from the nucellus.

In vitro nucellar embryogenesis has also been reported in *Aegle marmelos, Agrostemma githago, Malus domestica* and *Poncirus trifoliata*.

Studies on the nucellus culture have been extended to the species of other genera also. Mullins and Srinivasan (1976) succeeded in inducing nucellar embryony in ovule cultures of *Vitis vinifera*, a normally monoembryonate species. Unfertilized ovules cultured on Nitsch medium supplemented with benzyl adenine and β-NOA formed a nucellar callus, which subsequently produced embryoids. These embryoids produced plantlets when transferred to a semi-solid medium containing GA_3 and 2iP.

Rangaswamy (1982) cultured nucellus of several cucurbits – *Cocccinia, Cucumis, Lagenaria, Luffa* and *Trichosanthes*. The cultured nucelli produced profuse callus, but embryogensis was completely absent. Rangaswamy concludes that the lack of embryogenic capacity in nucellar cultures of Cucurbitaceae is due to a chemical block – cucurbitacins.

In unpollinated ovary cultures of *Cynanchum vincetoxicum*, Haccius and Hausner (1976) observed the development of green plantlets through the ovary wall. Histological studies proved them to be adventive embryos originating from the nucellar cells of unpollinated ovules. Some of these *in vitro* obtained nucellar plantlets were planted in the soil where they developed into flowering plants.

10.1.1. Culture Requirements and Technique

The MS medium is generally adequate for both polyembryogenic and monoembryogenic varieties. A high sucrose concentration of 5% is generally recommended. Complex substances, rich in amino acids, such as CH, malt extract or yeast extract should be supplemented along with MS medium. Sometimes, auxins are not essential for immature fruit tissues. Both, an auxin together with cytokinin, may stimulate embryogenesis.

1. Harvest immature green fruits with 2.5–3.0 cm diameter from the trees after 100–20 d of pollination.

2. Wash the fruits with tap water, then surface-sterilize by immersing them in commercial hypochlorite for 7–10 min. Include a few drops of Tween 20^{R} emulsifier in the disinfectant as a surfactant and rinse the fruits thrice in sterile distilled water.

3. Cut open the fruits under laminar flow hood, excise the ovules and transfer them to a sterile petri dish.

4. With the aid of a microscope using 30–60 X magnification, remove away the zygotic embryos along with the nucellus from the integument by piercing and pushing at the micropylar end. Discard zygotic embryos and place the nucellus on the media.

5. Callus formation from the nucellar cells takes place within 2–3 wk with a polyembryonal mass formed in 6 wk. Separate individual plantlets for subculture and eventually transfer to the pots. Indirect induction of adventive embryos occurs in polyembryonic varities.

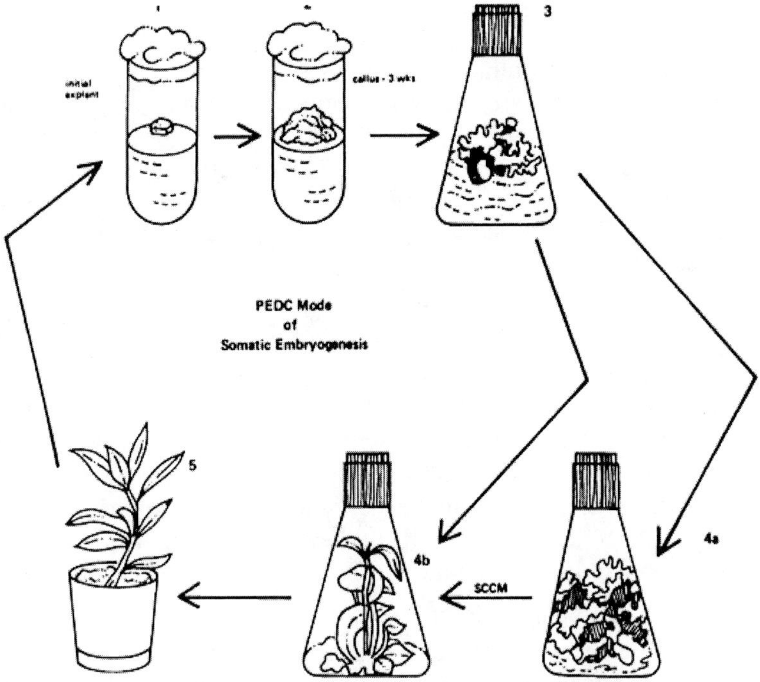

Fig. 32 Protocol for somatic embryogenesis in Citrus for PEDC mode of embryogenesis: 1. Callus formation from nucellar explants from immature fruits on MS medium with KN and IAA, 2. Subculture to a medium devoid of growth regulators and obtain polyembrony mass, 3. Subculture the proembryonal mass, 4a. Subculture embryo on to a growth-regulator-free medium, and 4b. Transfer the plantlets to the green house.

10.1.2. Applications

- The adventive embryos are of considerable importance to the horticulturists. They are generally uniform, and reproduce the characters of maternal plant without inheriting the variations brought about by the gametic fusion.
- Nucellar embryos from unpollinated ovules are also of considerable interest in the production of seedless varieties without the risk of introducing new viral diseases.
- It avoids the necessity of long periods of quarantine.

10.2. OVULE CULTURE

Ovule culture or seed culture may be described as the experimental system by which ovules are isolated aseptically from the ovary and are grown on a chemically-defined medium. White (1932), for the first time, cultured the ovules of *Antirrhinum majus* and reported callus formation from integument. La Rue (1942) cultured ovules of *Erythronium americanum* on WM fortified with IAA. Although there was increase in size, the seeds failed to mature.

However, the technique of ovule culture was developed and perfected by N. Maheswari and Lal (1958, 1961) at the University of Delhi. They reared the ovules of *Papaver somniferum* after fertilization with zygote or 2-celled proembryo and a few endosperm nuclei on WM fortified with KN. When the ovules were maintained in the culture for 50 d, they germinated. Sachar and Kapoor (1959) cultured the ovules of *Zephyranthes*, when they contained zygote and primary endosperm nucleus, on Nitsch medium supplemented with CM. The ovules showed mature embryo within 12 d and the seeds germinated *in situ* within 20 d.

Poddubnaja–Arnoldi (1959, 1960) cultured the ovules from pollinated ovaries of several orchids (*Calanthe veitcher, Cyprepedium insigne, Dendrobium nobile* and *Phalaenopsis schilleriana*) on simple 10% sucrose solution. Entry of the pollen tube into the ovule and development of the embryo has been observed in these cultures. Rangaswamy (1961) cultured the ovules of *Citrus microcarpa* and observed that the nucellar embryos showed development and differentiation on WM containing 5% sucrose. However, the embryos dissected from such ovules showed normal growth only on addition of CH (400 mg/l) to the medium.

Bajaj (1964) cultured the ovules of *Abelmoschus esculentus* 2 and 4 days after pollination (containing 4-celled proembryo with free nuclear endosperm and globular proembryo with cellular endosperm respectively), and observed that the ovules developed satisfactorily on the medium containing CH and CM. The ovules matured in 6 wk and also germinated.

Guha and Johri (1966) cultured the ovules of *Allium cepa* before and after pollination on Nitsch medium + IAA + GA_3 + 2% sucrose and observed that the ovules failed to develop. They further observed that the ovules cultured 13 d after pollination on Nitsch medium developed normally and germinated *in situ* after 20 d.

Wakizuka and Nakajima (1974) studied the *in vitro* development of ovules of *Petunia hybrida*. The ovules with early globular proembryo developed into mature seeds, on a medium supplemented with cucumber fruit-juice, adjusted to favourable osmotic value. They inferred that osmotic value of the medium influenced young ovules, which developed to maturity.

Chopra and Sabharwal (1963) cultured the pollinated ovules of *Gynandropsis*. They observed that the placental tissue had a beneficial effect on the growth and maturation of the seeds.

The effect of plant hormones on fibre and seed development in cotton has been studied by Beasley (1971) and Beasley *et al.* (1974). Fertilized ovules excised 2 d after anthesis grew and developed fibres only when floated on the liquid medium. Submerged ovules or ovules placed on identical, but agar medium, callussed or failed to grow. If cultured one day after anthesis, only a few ovules produced fibres, and none when cultured on the day of anthesis. Beasley and Ting (1973, 1974) studied the effect of growth substances on *in vitro* fibre development of cotton ovules. Both KN and ABA inhibited the development of the fibres. The immature ovules enlarged and produced fibres if the medium contained IAA or GA_3 or both.

In many interspecific and intergeneric crosses, the F_1 hybrid embryo frequently aborts in the developing seed, or the seeds produced are incapable of supporting the development of embryo. In their attempts on interspecific hybridization in *Abelmoschus*, Gadwal *et al.* (1968) observed that the hybrid embryo failed to develop beyond the heart or torpedo-shaped stage. They successfully obtained viable hybrids by employing both ovule and embryo culture in three out of five crosses attempted *viz.*, *Abelmoschus esculentus* X *A. ficulneus, A. esculentus* X *A. moschatus, A. tuberculatus* X *A. moschatus.*

Ovule culture assumes great significance in plant breeding. The hybrids, which usually fail to develop either due to abortion of embryo at an early stage or premature abscission of the fruits, could be obtained by ovule culture.

Recently, the more general term in ovule culture was introduced instead of ovary, ovule and fertilized ovule culture. Ovule culture is not likely to be useful where maternal tissue or endosperm exerts an inhibitory action on the development of embryo.

10.2.1. Applications

- In a normal process of hybrid production of certain crops, the hybrids fail to develop due to early embryo abortion and premature abscission of the fruits. Ovule culture can be used to rescue them.
- This technique omits excision of embryo by culturing the entire ovule, thus facilitates the time and effort involved.
- Ovule cultures are quite useful for studying early stages of dicot embryogenesis.
- Unfertilized ovule culture has further advantages than anther culture from the point of view of the embryogenesis efficiency, limited influence of donor plant genotype, and a higher frequency of haploids and dihaploid production.

10.3. OVARY CULTURE

Ovary culture may be described as the culturing of ovaries isolated from pollinated or unpollinated flowers. The concept of culturing excised flowers and floral parts *in vitro* originated with La Rue (1942) in a few angiosperms, and obtained root formation from the pedicels. Jansen and Bonner (1949) cultured the ovaries of *Lycopersicon pimpinellifolium* on a medium supplemented with casein (1951), *Nicotiana tabacum*, *Phaseolus vulgaris* and *Fragaria* sp.

Chopra (1958, 1962) studied the effect of IAA, IBA, KN and GA_3 on both unpollinated and pollinated ovaries of *Althaea rosea*. On Nitsch medium supplemented with IBA, the unpollinated ovaries developed into parthenocarpic fruits, which are comparable in size with those that developed *in vivo*. Although the addition of KN or IAA alone had no pronounced effect, KN and IAA used in conjunction acted synergistically in inducing parthenocarpy. In culture of pollinated ovaries, maximal growth of the fruit occurred on Nitsch medium + IBA in darkness; 50% of the fruits matured in 25–30 d. Chopra (1958, 1962) also observed that the growth of ovaries was considerably influenced by the presence of calyx. If the calyx was removed prior to culture, the fruits grew to a diameter of only 12 mm, the endosperm remained free nuclear and the embryo reached only heart-shaped stage. If the calyx was left intact, the fruits enlarged in diameter of 19 mm, the endosperm became cellular, and the embryo showed normal development.

Similarly, in *Iberis amara* (N. Maheshwari and Lal, 1961) flowers cultured 1 d after pollination produced normal fruits; only if calyx was not removed before culture. If the calyx was removed, addition of sucrose (5%) was necessary to obtain satisfactory growth.

Puri (1963) studied the effect of growth substances such as CH, IAA, CM and yeast extract on the growth of embryo in the culture of ovaries, flowers and portion of spikes of *Aerva tomentosa*, an obligate apomict. On the other hand, only 7% of the ovules developed into seeds with poorly-differentiated embryos in ovary culture; and nearly 15% of the seeds bore mature embryos in flower cultures. The addition of yeast extract, CH or CM increased the percentage of seed-set only slightly. The best response was, however, elicited when portions of the spike were cultured. In the presence of IAA and yeast extract, the seed-set was comparable to that in nature, and the growth of the embryo and endosperm was quite normal.

Guha and Johri (1966) studied the minimum requirement of developing fruits of *Allium cepa*. The maximum growth of fruit occurred on Nitsch medium supplemented with GA, IAA and KN while only a few viable seeds were produced on the BM. The addition of IAA or tryptophan increased seed-set to 20% and 30% respectively.

Ovary culture has been successfully employed to obtain hybrids of diploid *Brassica chinensis* and autotetraploid *B. pekinensis* (Inomata, 1968), which are normally cross-incompatible. The ovaries excised 4 d after pollination were cultured on Nitsch medium supplemented with various growth substances such as CH, GA_3, IAA, KN, yeast extract, tomato juice and CM. Hybrid seeds were produced on a medium supplemented with yeast extract and CH. The triploid hybrid plants, thus obtained, showed intermediate characters of their parents.

10.3.1. Advantages of Ovary Culture

- It is successful in inducing polyembryony in various parts of the plant.
- It is used in the fundamental and applied aspects of development of the fruits and seeds.
- It provides important information about the role of accessory floral parts in the fruit development.
- In certain plants, apomixis occurs in which seed development takes place without fertilization. In these apomictic plants,

pollination alone stimulates the growth of ovary and development of the seeds. In such plants, ovary culture is useful in the study of stimulus provided by pollination.

- Test tube fertilization and ovary or pistil culture has been used to circumvent various problems associated with hybridization such as pollen germination on the stigma, insufficient growth of the pollen tube and self incompatabilty or sterility.

10.4. FLOWER CULTURE

The technique of culturing young, excised floral primordia affords a potentially-useful tool for any study concerned with the control of flower morphogenesis. It permits assessment of the influence of growth regulators and nutrients on buds, without presence of intervening vegetative tissues. Secondly, one can test whether the floral apex behaves autonomously i.e., whether the sequence of organ initiation and subsequent growth is regulated from within the isolated immature bud. Lastly, it facilitates easy handling and allows a variety of surgical manipulations of the excised buds in culture.

The technique of floral bud culture was first initiated by La Rue (1942) who cultured pollinated flowers of *Caltha*, *Kalanchoe* and *Lycopersicon*. Galun *et al.* (1962, 1963) achieved success in culturing detached floral buds of *Cucumis sativus*. If left *in situ,* all the floral buds would have developed into male flowers, depending on the nodes on which they were borne. However, the excised buds were induced into female flowers on modified WM supplemented with IAA. The GA_3 counteracted this effect, but had no stimulatory effect on its own.

Topfer *et al.* (1963, 1966) grew the floral buds of *Aquilegia formosa* to maturity. Blake (1966) first reported complete development of the excised buds into flowers in *Viscaria candida* and *V. cardinalis.* In these species, excised floral apices were raised on MS medium supplemented with CM, CH, inositol, amino acids, GA and IAA.

In *Begonia francois*, a monoecious species, normal flower buds developed when young floral primordia were cultured on a medium containing BA. Gibberellins promoted the initiation of floral organs.

In *Ranunculus scleratus* (Konar and Nataraja, 1964, 1969), the buds comprising the primordia of various organs failed to complete the normal development on any of the nutrient media tested, but they all formed callus, which subsequently differentiated into roots, shoots and embryoids leading to the formation of plantlets. The cut end of the undifferentiated flower buds proliferated rapidly resulting in a mass of white, friable callus. Optimal callus formation was observed on WM containing 2% sucrose, 10% CM, 500 ppm CH and 1 ppm 2, 4-D.

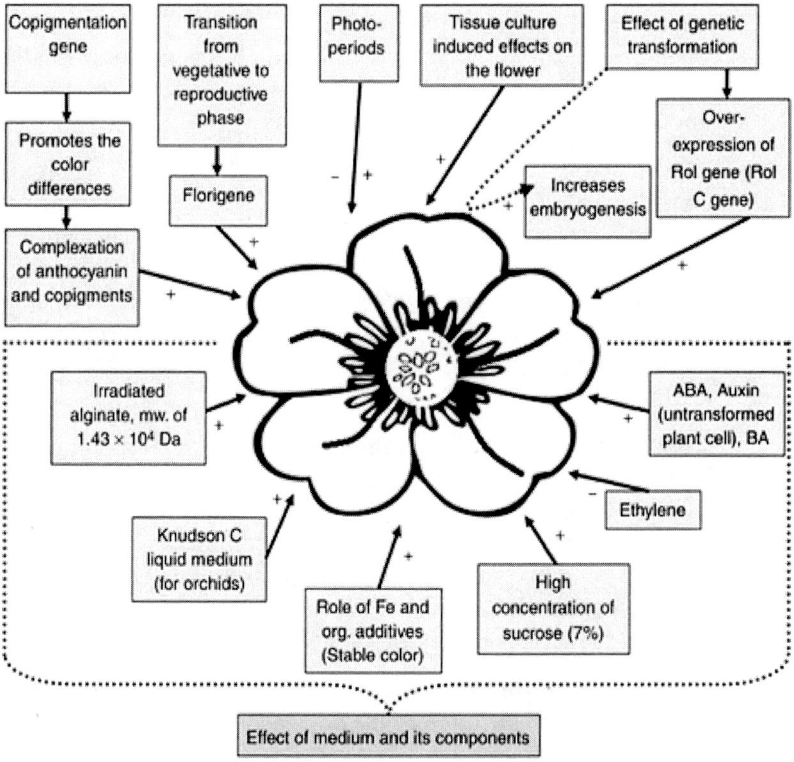

Fig. 33 Effect of factors on flowering

Flowering is dependent on the genes that act to promote flowering. In addition, photoperiods, vernalization, role of carbohydrates and phytohormones also play an important role in the promotion of flowering. There are several factors that either stimulate or inhibit the flowering *in vitro:*

- **Effect of gene transformation:** Traditionally, new traits have been introduced into ornamental plants through classical breeding. Genetic modification always plays an important role in floriculture industry. Genes, for example rol genes, have potential utility in the floriculture industry. Molecular mechanisms underlying phenotypic expression of these genes are of great importance to the industry. Rol genes of the Ri plasmid of *Agrobacterium rhizogenes* in plants alter the developmental processes of several plants and affect their architecture. Improvement of the ornamental and horticultural traits can be done by introduction of rol C genes. More specifically, the rol C gene is responsible for inducing smaller flowers and advanced flowering, and therefore, offers useful tools for improving ornamental flowers.

- **Regeneration potential of transformed explants by auxin supplementation:** Auxin supplementation is reported to regenerate two to three times more flower buds than explants from untransformed tobacco. Tobacco plant cell pRi 1855 transformation by *Agrobacterium rhizogenes* makes the plant cell insensitive to supplemented ethylene concentration, which inhibits auxin promotion and flower bud formation in its explants.

- **Genes that promote copigmentation:** Colour differences in the plants can be made by introducing a gene that promotes copigmentation. In this process, anthocyanins and copigments such as flavones or flavonols form complexes that lead to shifting of the visible absorption spectrum of the flowers so that it becomes longer and flowers look blue. In *Torenia*, DFR gene introduced to produce a blue trait in flowers.

- **Stable colouration in flowers:** Tissue culture induced flower-colour changes in *Saintpaulia* due to excision of the F3'5'H promoter.

- **Increased production of embryogenic calli by transformation:** When compared with the effect of controlled medium, transformation of tobacco arabinogalactan protein genes significantly increased the production of embryogenic calli from banana immature male flowers.

- **Florigen:** It is now widely accepted that a leaf-borne signal, florigen moves via the phloem from the leaves to the shoot apical meristem to trigger its reprogramming to produce flowers. Maize FT homologue ZCN8 has florigenic function, which plays a pleiotropic role in the regulation of generalized growth of vegetative and reproductive tissues (Danilevskaya *et al.,* 2011).

- **Effect of hormones:** *Dendrobium candidum* requires 3–4 y of cultivation before flowering. Supplementing BA or combination of NAA and BA promotes flowering within 3–6 m with 31–45% frequency. However, later it was found that the pretreatment of protocorms, initially with ABA and later transferring to MS medium with BA, enhanced the flowering to 82%.
- **Effect of sugars:** Sucrose concentration plays an important role and its absence may also affect the colour defect in some plants. Nagira *et al.* (2006) reported that 7% sucrose elevated endogenous ABA levels, and 1.5% sucrose resulted in decrease in the ABA level and caused failure of induction of anthocyanin synthesis.

10.4.1. Advantages

In vitro flowering overcomes the problems associated with flower growth and developments as well as fruit and seed production *in vitro.* Hence, it is helpful in conservation and continuous supply of the plant material throughout the year by understanding its flowering behaviour *in vitro*.

■■■■

11 EMBRYO AND ENDOSPERM CULTURE

11.1. EMBRYO CULTURE

The first systematic attempt to culture the embryos of angiosperms was made by Hannig (1904). He cultured mature embryos of *Raphanus caudatus*, *R. landra*, *R. sativus* and *Cochlearia danica* (Family: Brassicaceae) in mineral salts medium supplemented with sugar, and obtained transplantable seedlings. While working with embryos from a wide range of families and genera, Dietrich (1924) made two generalizations (a) the embryos, grown *in vitro*, usually skipped a rest period and germinated, (b) the embryos taken into culture at less than one-third of mature size were unable to produce viable plants capable of reaching maturity. Embryos from immature seeds tended to form malformed seedlings. Later on, many workers obtained plants by culturing excised embryos. Laibach (1925, 1929) cultured excised embryos from seeds of an interspecific cross *Linum perenne* X *L. austrianum* and succeeded in raising hybrid plants. The seeds formed after crossing shrivelled, became very light, incapable of germination. Laibach excised embryos from such seeds and cultured them on a moist filter paper or on cotton wadding containing glucose or sucrose, and was able to raise the hybrids.

Embryo culture can be categorized into two types:

1. Culture of immature embryos and of mature embryos
2. Immature embryos originating from unripe seeds and hybrid seeds fail to germinate.

Excising such embryos is difficult and they require a complex nutrient medium. Mature embryos are excised and cultured to avoid inhibition in the seed for germination and they require simple nutrient medium.

Mature embryo cultures are carried out in the following conditions:

- When the embryos remain dormant for long periods
- Low survival of embryos *in vivo*
- To avoid inhibition in the seed for germination
- To convert sterile seeds to viable seedlings

Generally, the ovule or ovaries are surface-sterilized and the embryo is excised under aseptic environment. Media constituents of *in vitro* growth of young or immature embryos differ from those of mature embryos. This, often necessitates, the transfer of embryos from one medium to another in order to achieve their full development.

While carrying out *in vitro* culture studies of embryos in *Datura* van Overbeek *et al.* (1941, 1942) has made the following observations:

- Mature embryos form normal seedlings on mineral salts and dextrose.
- Young torpedo-shaped embryos (up to 0.5 mm) require a medium containing minerals, dextrose, glycine, thiamine, ascorbic acid, nicotinic acid, Vitamin B_{12}, adenine, succinic acid and pantothenic acid.
- Embryos smaller than 0.5 mm in length could be cultured successfully only upon the addition of CM to the medium required by torpedo-shaped embryos.

Addition of CM in the medium has proved to be beneficial for the development of embryo. In barley, undifferentiated embryos were successfully cultured by supplementing the medium with 20% CM.

Osmotic value of the nutrient medium seems to be an important factor for the proper growth of embryo. Isolated mature embryos grow fairly well with 2% sucrose in the culture medium, but younger embryos require higher levels of carbohydrate for normal embryo development. Raghavan and Torrey (1969) had noted a stimulatory effect of high concentration (12–18%) of sucrose on *in vitro* development of excised globular embryos of *Capsella bursa-pastoris*.

11.1.1. Technique

1. Take out the hybrid embryo from the ovule in which endosperm development has failed by excision. Choose

another normally-developed ovule with endosperm enclosing an embryo. Dissect this ovule and press out the normal embryo. This leaves a normal endosperm with an exit hole.

2. Now, insert the hybrid embryo into the normal endosperm through exit hole. This results in embryo-endosperm transplant. Culture this transplant in a suitable medium. Many interspecific and intergeneric plants have been raised by using embryo-endosperm transplant e.g., hybrid plants of legumes.

11.1.2. Nutritional Requirements of Embryo Cultures

There are two phases in the embryo development and the nutritional requirement varies accordingly:

1. **Heterotrophic phase:** This is an early phase, and the embryo is mostly dependent on the endosperm and maternal tissues for nutrient supply.
2. **Autotrophic phase:** This phase is characterized by the metabolic capability of the embryo to synthesize substances required for its growth, which slowly makes it independent. The critical stage is the intervening phase between the heterotrophic and autotrophic phases. The nutrient supply is highly variable at this phase, which mostly depends on the plant species.

In general, the composition of the medium for culturing immature embryos is more complex than that required by mature embryos, which can grow on a simple inorganic medium. Further, the transfer of embryos from one medium to another is frequently needed in order to achieve full development of the embryos. Some salient features of the medium and culture conditions are listed below:

- Inorganic constituents of MS, B5 or WM are adequate.
- Sucrose is most commonly-used energy source.
- Ammonium nitrate is the preferred source of nitrogen.
- The CH, rich in various amino acids, is frequently used.
- Certain natural plant extracts with embryo factor promote embryo cultures e.g., liquid endosperm of CM. The embryo factor is believed to supply certain amino acids, sugars, growth regulators etc.

- In general, growth regulators are not required as they induce callus formation.
- Embryos grow well in the pH range of 5–7.5.
- An incubation temperature of 24–26°C is ideal.
- Better growth of the embryos is observed in darkness; the embryos are then transferred to light for germination.

During the culture conditions, the embryos are grown into plantlets, and then transferred to sterile soil for full-fledged growth to maturity.

11.1.3. Embryo-Nurse Endosperm Transplant (Embryo Rescue)

One of the main problems of hybridization is rescuing the hybrid embryo, which normally aborts at a very early stage of development. Kreuse (1974) implanted immature embryos arising in a cross *Hordeum* X *Secale* on cultured barley endosperm and recorded significant improvement.

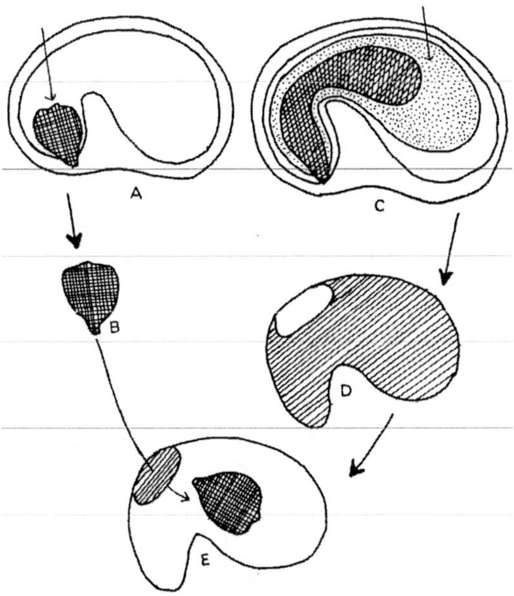

Fig. 34 Endosperm transplant culture of hybrid embryo (after Williams and De Latour, 1989)

De Latour *et al.* (1978), Williams and De Latour (1980) described embryo rescue method (Fig. 34). They inserted an excised hybrid embryo into cellular endosperm dissected from a normally-developing ovule of one of the parents or a third species, and cultured the nurse endosperm with the transplanted embryo on the nutrient medium. They obtained interspecific and intergeneric hybrid plants of forage legumes – *Trifolium*, *Lotus* and *Ornithopus* – that could not be reared by culturing embryos on the medium directly.

11.1.4. Practical Applications

Obtaining hybrids

The culture of excised embryos can be advantageously exploited by the plant breeder. Very often interspecific crosses fail due to arrested development of embryo either due to abortion of embryo at an early stage or due to degradation of endosperm. A number of hybrids have been successfully raised through embryo culture. These include *Abelmoschus esculentus* X *A. manihot*, *A. esculentus* X *A. moschatus*, *A. tuberculatus* X *A. moschatus* (Gadwal *et al.,* 1968), *Hordeum vulgare* X *Secale cereale* (Kreuse, 1974), *Hordeum vulgare* X *Agropyron repens*, *Hordeum vulgare* X *Triticum* sp., *Lycopersicon esculentum* X *L. peruvianum*, *Brassica chinensis* X *B. pekinensis*.

Overcoming dormancy and shortening breeding cycle

In horticultural crops, breeding work is delayed due to long periods of dormancy in the seeds. In these plants, breeding cycle can be shortened using embryo culture technique. For example, using this technique in *Iris,* Randolph and Cox (1943) could shorten the life cycle from 2–3 y to less than 1 y. Similarly, in *Rosa,* it is possible to obtain two generations of flowering by culturing embryos. Overcoming seed dormancy and shortening of their breeding cycles have also been reported in weeping crabapple (*Malus* sp.), hollies (*Ilex*), American basswood (*Tilia americana*) etc.

Overcoming seed sterility

In early-ripening fruit cultivar, seeds do not germinate because their embryos are still immature. Using the embryo culture method it is possible to raise the seedlings from sterile seeds of early-ripening stone fruits (sweet cherry), peach, apricot and plum.

'Makapuno' coconuts develop soft, solid, fatty endosperm in place of liquid endosperm. Such nuts are very rare and expensive. Under normal conditions, such coconut seeds fail to germinate. De Guzman *et al.* (1971) excised such embryos and cultured them, and raised plantlets. About 85% of the nuts borne by these plants were Makapuno.

Similar procedure can be applied for *Musa balbisiana, Maranta leuconeura, Colocasia esculenta* and *Pinus armandii* X *P. koriensis.*

Production of monoploids

Embryo culture has been used in the production of monoploids (haploids) of barley. With the cross *Hordeum vulgare* X *H. bulbosum* fertilization occurs normally, but thereafter, chromosomes of *H. bulbosum* are eliminated resulting in a monoploid *H. vulgare* embryo. Such monoploid embryo can be rescued through embryo culture (Kasha and Kao, 1970). Following the bulbosum technique, monoploids were obtained from a cross *Agropyron tsukushiense* X *Hordeum vulgare* (see Kott and Kasha, 1985).

Rapid seed-viability test

The possibility of breaking seed dormancy by embryo culture also allows the use of this technique as a rapid test for seed viability. Barton (1961) has suggested that germination of excised embryos is reliable and capable of more exact interpretation than the commonly-used staining tests for seed viability.

Clonal micropropagation

The regenerative potential is an essential prerequisite in non-conventional methods of plant genetic manipulation. Due to their juvenile nature, embryos have a high potential for regeneration, and hence, may be used for *in vitro* clonal propagation. This is essentially true in case of grasses. Both organogenesis and somatic embryogenesis have been induced in major cereals and forage grasses from embryoic tissues. Generally, callus derived from immature embryos of cereals has the desired morphological potential for regeneration and clonal propagation.

11.2. ENDOSPERM CULTURE

Lampe and Mills (1933) made the first attempt to culture the endosperm. They grew the immature endosperm tissue of maize on a medium containing extract of potato or young corn kernels and observed slight proliferation of endosperm layers around the embryo. La Rue (1949), for the first time, succeeded in establishing a continuously-growing tissue from immature endosperm of maize. Subsequently, tissue cultures were established from the immature endosperm of *Asimina triloba* (Lampton, 1952), *Lolium perenne* (Norstog, 1956) and *Cucumis sativus* (Nakajima, 1962). In all these cases, endosperm grows in the cultures only when it is excised 7–10 d after pollination. There are no reports on proliferation of free nuclear endosperm. Nakano *et al.* (1975) pointed out that in rice, only the cellular endosperm proliferated. In maize, the endosperm becomes cellular 3–5 d after pollination, and the cells remain meristematic for 8 d. Later, the mitoses are restricted to only the outermost layer.

In most cereals, the mature endosperm does not respond to the cultural conditions. Even endosperm of some dicots (*Cucumis sativus*) exhibits similar behaviour.

In contrast to this, the mature endosperms of Euphorbiaceae, Santalaceae and Loranthaceae respond favourably to the culture treatments. Association of embryo tissue in the initial stages seems essential for inducing proliferation of the mature endosperm in the members of Euphorbiaceae and Santalaceae. The embryo can be removed as soon as proliferation of the endosperm has started. Presumably, the embryo releases some factor(s) required to activate the cells to divide. In some species such as *Croton bonplandianum* and *Putranjiva roxburghii,* the embryo factor(s) may be replaced by GA_3.

Time required for initiation of proliferation was 10 d for *Ricinus*, 10–12 d for *Jatropha*, 21 d for *Putranjiva* and 21–25 d for *Santalum*. Pre-soaking of the endosperm in GA_3 (2 ppm) could reduce the period from 10–7 d in *Ricinus*. The parasitic angiosperms tend to bypass callus phase and undergo organogenesis directly from the cultured endosperm.

Direct differentiation of the shoots and haustoria are reported from the mature endosperm of semi-parasitic angiosperms such as *Exocarpus*, *Taxillus*, *Leptomeria*, *Scurrula*, *Dendrophthoe* and *Santalum*. In all these cases, shoot buds were formed after negligible callusing or without callusing. However, in autotrophic taxa, differentiation of the shoot buds or roots was through callus only.

The BM of WM or MS medium is generally used in the endosperm culture. The BM has been supplemented with tomato juice, yeast extract, grape juice and maize grain juice to induce callus from immature endosperm. Only yeast extract was effective for *Lolium* endosperm.

The maiden report of cellular totipotency of the endosperm tissue was that of Johri and Bhojwani (1965) in *Exocarpus cupressiformis*. They observed the development of eight shoot buds in a single explant (seed) cultured on WM supplemented with IAA, KN and CH. These buds, which are triploid, originated from the peripheral cells of the endosperm, and possessed distinct shoot apices and well-differentiated vasculature. Since then, organogenesis has been induced in the endosperm cultures of apple (Mu *et al.,* 1971), *Citrus grandis* (Wang and Chang, 1978), *Dendrophthoe falcata* (Nag and Johri, 1971), *Scurrula pulverulenta* (Bhojwani and Johri, 1970), sandalwood (Lakshmi Sita *et al.,* 1980), *Putranjiva roxburghii* (Srivastava, 1973), *Jatropha panduraefolia* (Srivastava, 1971), *Codiaeum variegatum* (Chikkannaiah and Gayathri, 1974), *Petroselenium hortense* (Masuda *et al.,* 1977), *Annona squamosa* (Nair *et al.,* 1986), *Actinidia chinensis* (Gui *et al.,* 1982), *Mallotus philippensis* (Sehgal and Abbas, 1996), *Emblica officinalis* (Sehgal and Sunila, 1985), *Actinidia* hybrids (Mu *et al.,* 1990), Maize (Zhu *et al.,* 1988), rice (Nakano *et al.,* 1975; Bajaj *et al.,* 1980), and *Hordeum vulgare* (Sun and Chu, 1981).

11.2.1. Factors Affecting Shoot-bud Differentiation

The endosperm may form buds either directly or it may first proliferate into a callus mass followed by organogenesis. An exogenous cytokinin is always necessary for a shoot-bud differentiation from endosperm tissue. In 'seed' cultures of *Scurrula pulverulenta* on a medium supplemented with IAA, KN and CH, 15% of the explants showed surface callusing of the endosperm. In another 13% cultures, green buds developed from the unproliferated regions of endosperm. Omission of CH from the medium promoted shoot-bud differentiation to 26%. Further studies revealed that KN alone was sufficient to induce shoot-bud differentiation; CH and IAA counteracted the effect of KN. However, in *Petroselenium,* immature endosperm showed embryogenesis on hormone-free MS medium (Masuda *et al.,* 1977). The morphogenic response of mature endosperm of *Taxillus* is affected by the presence of CH.

Various physical factors also have an influence on *in vitro* growth of endosperm tissue. In *Asimina*, Lampton (1952) observed increase in the growth of endosperm callus at pH 4.0 and at pH 5.0 (95%). Likewise, pronounced growth of *Ricinus* endosperm at pH 5.0 and of *Jatropha* and *Putranjiva* at pH 5.6 were noticed. In general, pH 5.5–5.8 seems to support the best growth of the endosperm tissue in cultures.

The optimum temperature range supportive of endosperm culture is about 24–27°C. In maize, there was a decrease in growth by 50% at 30°C and four-fold increase at 25°C as compared to the growth at 20°C. Cultures maintained only in dark showed satisfactory response. *Jatropha* and *Ricinus* exhibited best response at 24–26°C. Mostly, 12–16 h diffuse-daylight period supported callusing as well as regeneration.

Plantlet formation from endosperm callus may follow the embryogenic mode of development. Embryo differentiation occurs when the proliferated tissue is transferred from a callusing medium to a BM with or without gibberellin.

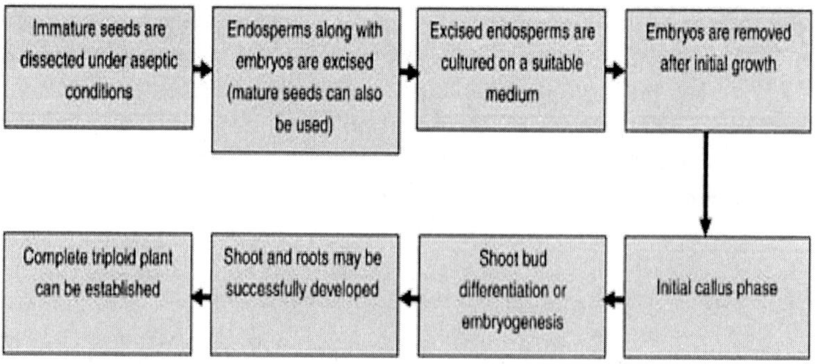

Fig. 35 Steps involved in endosperm culture technique

11.2.2. Application of Triploids in Plant Improvement

The technique of endosperm culture enabled the production of triploid plants. Triploid plants are self-sterile and usually seedless. This character increases the edibility of fruits and is desirable in such plants as apple, banana, grape, mango, watermelon etc. for their edible fruits. Seed sterility in timber, fuelwood or latex-yielding plants is not a serious setback because their vegetative parts are important and these plants can be multiplied vegetatively. In such plants, triploids show better performance than their relative diploids or tetraploids. For example, the triploids of *Populus tremuloides* have better-quality pulp wood, which is important to the forest industry.

■■■■

12 PROTOPLAST CULTURE AND SOMATIC HYBRIDIZATION

12.1. INTRODUCTION

In eukaryotes, the transfer of genetic material from one individual to another is conventionally achieved through sexual breeding, the scope of which is extremely limited, particularly in animals. Even in plants, where fairly-distant species could be crossed, it has not always been possible to obtain full hybrids between the desired individuals because of the sexual incompatibility barriers. This has often proved a serious handicap in crop improvement programmes through hybridization. In this respect, cell fusion offers a novel approach to distant hybridization (somatic hybridization). Fusion of the cells, whether in plants or animals, must occur through the plasma membrane. Unlike animals, in plants, the plasma membrane is bound by a rigid cellulose wall, and the adjacent cells are cemented together by a pectin-rich matrix. It is mainly for this reason that the somatic cell genetics is more advanced with animal than with the plant systems. It is only since 1960, when E.C. Cocking demonstrated the feasibility of enzymatic degradation of plant cell walls to obtain large quantities of viable naked cells, hereafter called protoplasts that real interest in genetic modification of somatic cells in higher plants has developed. Actually, active contributions in this area started appearing after 1970.

Plant cells without cell walls are referred to as protoplasts. The non-conventional genetic procedure involving fusion between isolated somatic protoplasts under *in vitro* conditions and subsequent development of their product to a hybrid plant is known as somatic hybridization or parasexual hybridization.

12.2. SOURCES OF PROTOPLASTS

- **Leaves:** The leaf is the most convenient and popular source of the plant protoplasts because it allows isolation of a large number of relatively-uniform cells. Sometimes, it is difficult to peel the epidermis from the leaves of some species (monocots). In such cases, the leaf material is cut into small pieces (about 1 mm^2) and then combined with vacuum infiltration. This procedure allows adequate infiltration of the enzymes into the leaf cells. As soon as the vacuum is removed, the leaf pieces will sink and eventually release the mesophyll protoplasts. Through this approach, a large yield of protoplasts can be obtained in comparatively short periods.

- **Callus cultures:** Young callus cultures are also ideal material for obtaining large quantities of protoplasts. Older cultures tend to form giant cells with thick cell walls, which are difficult to digest; therefore, young actively-growing callus is subcultured and used after 2 wk for protoplast isolation.

- **Cell suspension cultures:** Actively-growing young suspensions are ideal material for the isolation of protoplasts in large quantities. A high-density cell suspension is centrifuged. After removing the supernatant, cells are incubated in an enzyme mixture (cellulase + pectinase) in a culture flask placed on a platform shaker for 6 h to overnight depending on the concentration of the enzymes.

- **Pollen protoplasts:** Protoplasts can also be obtained from pollen. Microspores, in various stages of development, are collected under sterile conditions, and incubated in a cocktail of cellulase (2%), macerozyme (1%) and helicase (0.5%), with potassium dextran sulphate (0.1%) in 10–15% sucrose for 24 h period, and changed to fresh enzyme mixture after 12 h. Gentle agitation facilitates protoplast release.

- Protoplasts can also be obtained from pollen mother cells, pollen tetrads and mature pollen grains also.

- **Seedlings:** Protoplasts can also be obtained from root and stem of the young seedling.

12.3. ISOLATION OF PROTOPLASTS

Methods of protoplast isolation can be classified into three main groups: (1) Mechanical (non-enzymatic), (2) Sequential enzymatic (two-step), and (3) Mixed enzymatic (simultaneous) procedures.

12.3.1. Mechanical Method

The isolation of protoplasts from higher plants was pioneered by Klercker in 1892. The procedure followed by him was largely mechanical; the cells were kept in a suitable plasmolyticum and cut with a fine knife. In this process, some of the plasmolysed cells were cut only

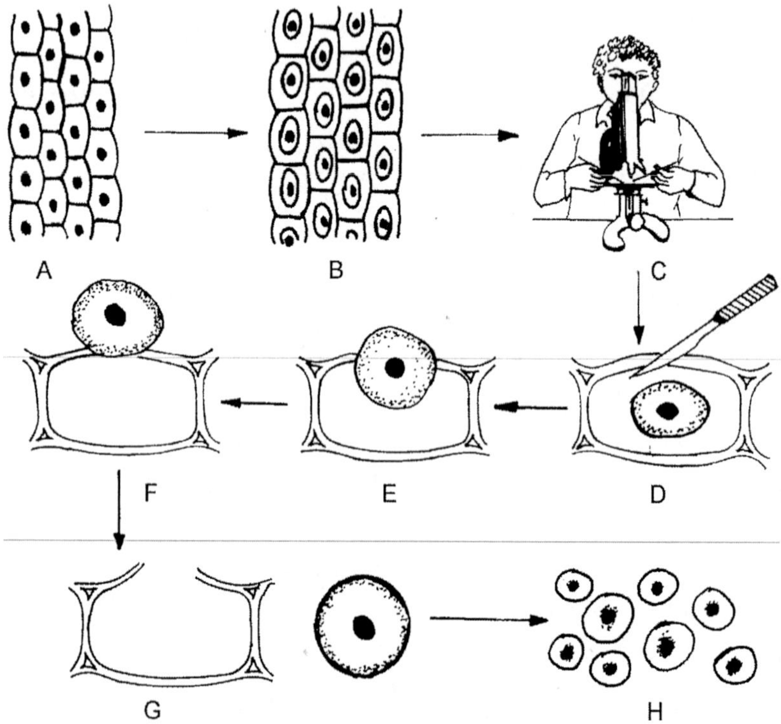

Fig. 36 Method of mechanical isolation of protoplasts: A. A small piece of plant tissue, B. Plasmolysis of cells, C-D. Cutting of cell wall by microscalpel under microscope, E-F. Subsequent stages of liberations of protoplasts, G. Isolated protoplast and empty cell, and H. Isolated protoplasts.

through the cell wall, releasing protoplasts (Fig. 36). This procedure for protoplast isolation is applicable only to vacuolated cells of storage tissues (onion bulbs, scales, radish root, mesocarp of cucumber and beetroot). Another limitation of this mechanical method is that the yields are extremely low.

12.3.2. Enzymatic Method

The somatic cells of the higher plants are bound by a rigid cellulose cell wall, and the adjacent cells are cemented together by a pectin-rich matrix. Protoplasts from such cells can be isolated by enzymatic method. Pectinases and cellulases are used for large-scale isolation of the protoplasts. These enzymes can be used: (1) sequentially (two-step) or (2) simultaneously (one-step). The concentration of the enzymes and the duration varies with the tissue. Cocking (1960) for the first time demonstrated the large-scale isolation of the protoplasts from root tips of tomato (*Lycopersicon esculentum*) by using concentrated solution of cellulase enzyme (prepared from the cultures of the fungus *Myrothecium verrucaria*). Further progress in the enzymatic isolation of protoplasts was achieved as soon as cellulase and macerozyme (pectinase) enzymes became available commercially in 1968. A range of enzyme preparations are now available commercially (Table 10).

Table 10: Some commonly-used commercially-available enzymes for protoplast isolation

Enzyme	Source	Supplier
Cellulases		
Onozuka RS	*Trichoderma viride*	Yakult Honsha, Japan
Cellulase R-10	*Trichoderma viride*	Yakult Honsha, Japan
Cellulysine	*Trichoderma viride*	Calbiochem, USA
Driselase	*Irpex lactes*	Kyowa Hakko Kogyo, Japan
Meicelase-P	*Trichoderma viride*	Meiji Seika Kaisha, Japan
Hemicellulases		
Hemicellulase	*Aspergillus niger*	Sigma, USA
Rhozyme HP-150	*Aspergillus niger*	Rohm and Hass, USA
Zymolyase	*Arthrobactor luteus*	Sigma, USA
Pectinases		
Macerozyme R-10	*Rhizopus* sp.	Yakult Honsha, Japan

Enzyme	Source	Supplier
Macerase	*Rhizopus* sp.	Calbiochem, USA
Pectinase (purified)	*Aspergillus niger*	Sigma, USA
Pectolyase Y23	*Aspergillus japonicus*	Seishin Pharmaceutical, Japan
Pectinol	*Aspergillus niger*	Rohm and Hass, USA

Takebe *et al.* (1968) employed sequential or two-step procedure for isolating mesophyll protoplasts in tobacco using commercial preparation of the enzymes.

Essentially, the technique involves the following steps:

1. Surface-sterilize the leaves.
2. Peel off the lower epidermis.
3. Give pre-enzymatic treatment.
4. Incubate the stripped leaf tissue in an enzymatic solution.
5. Isolate the protoplasts.

Fully-expanded leaves are excised from 7–8 wk old plants growing in a greenhouse. The leaves are surface-sterilized by first immersing in 70% ethanol for 30 s followed by rinsing in 0.4–0.5% sodium hypochlorite solution for about 30 min. The leaves are then repeatedly washed in sterile distilled water. The lower epidermis of the sterilized leaves is carefully peeled with the help of fine forceps and the stripped leaves are cut into small pieces of 4 cm^2. Peeling the epidermis would be relatively easy if the sterilized leaves were allowed to become flaccid, or if the water supply to the plants is limited before the leaves are excised. Protoplasts are isolated by two-step enzymatic treatment from the peeled leaf tissues. First, stripped leaf pieces are placed in 20 ml enzyme medium (0.5% macerozyme, 0.8 M D-mannitol and 0.3% of potassium dextran sulphate at pH 5.8) in 100 ml flasks and gently shaken in a water bath at 25°C by a reciprocal shaker. Fresh medium is replenished at 30 min intervals for 2 h. The mesophyll cells, thus isolated by maceration, are washed repeatedly with 0.8 M mannitol containing 0.1 M calcium chloride and suspended in 2% solution of cellulase (prepared from fungus *Trichoderma viride*) in 0.8 M mannitol at pH 5.4. The suspension is incubated at 36°C for 2–3 h with occasional shaking. The cellulase digests the cell walls resulting in the release of protoplasts (Fig. 37).

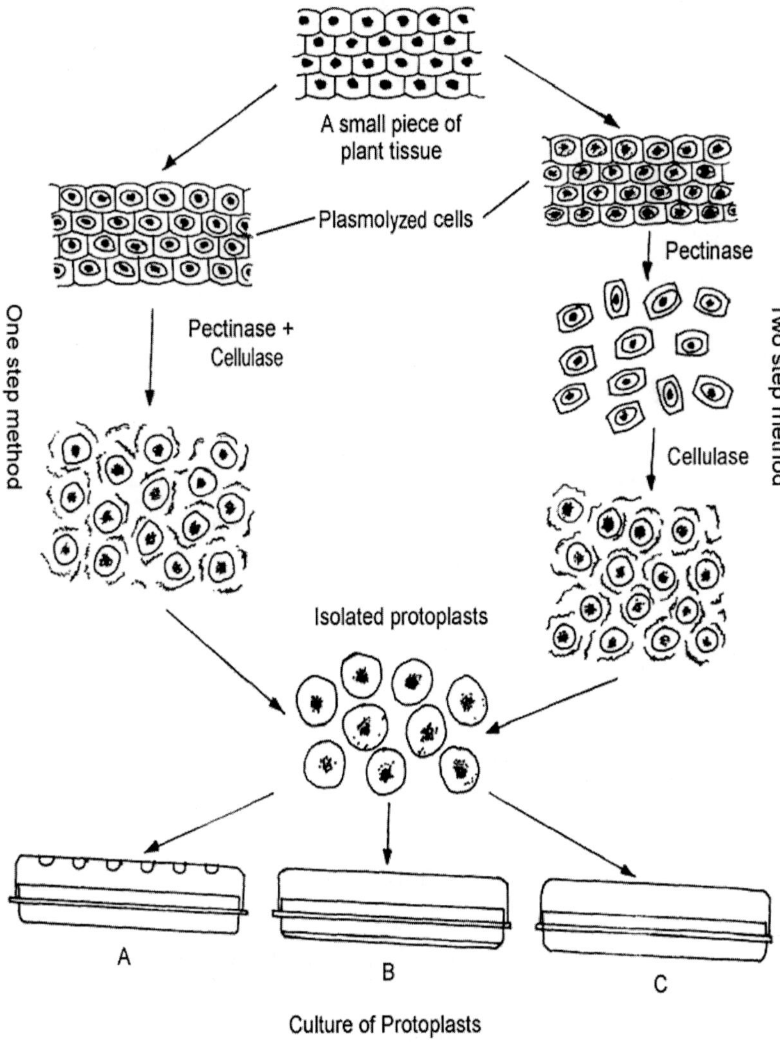

Fig. 37 Methods of enzymatic isolation of protoplasts and their culture; A. Hanging droplets method of culture, B. Co-culture, and C. Plating of protoplasts

Fully expanded leaves

↓

Surface sterilize

70% Ethanol 30 s

0.4–0.5% Sodium hypochlorite 30 s

↓

Peel-off the lower epidermis

↓

Cut into pieces of 4 cm^2

↓

Macerate for 2 h at 25°C

in a mixture containing

↓

0.5% Macerozyme,

0.8 M Mannitol,

0.3% Potassium dextran sulphate.

Change the mixture every 30 min

↓

Wash thoroughly the isolated mesophyll cells in

0.8 M solution of mannitol

↓

Degrade cell walls by incubating the free
cells for 2 h at 36°C in a mixture containing

2% Cellulase,

0.8 M Mannitol

↓

Wash the released protoplasts with 0.8 M mannitol

↓

Culture the protoplasts

or

Fuse them for somatic hybridization

Fig. 38: Summarized procedure for isolation of protoplasts from mesophyll cells of tobacco

Power and Cocking (1968) mixed two enzymes together (simultaneous procedure) and isolated protoplasts in one-step. The peeled leaf tissues were placed directly in an enzymatic mixture containing 0.5% macerozyme and 2% cellulase in 0.7 M sorbitol or mannitol solution. About 1 g of leaf material was added to 10 ml of the enzyme mixture (pH 5.4) incubated at 25°C for 4–6 h in darkness. The leaf pieces were gently squeezed with a Pasteur pipette to liberate protoplasts. The leaf debris was removed by passing through a 60–80 μm mesh sieve. The filtrate was transferred to the tubes and centrifuged at least thrice at 100 g for 1 min. Each time the protoplasts sank to the bottom; the supernatant enzyme and the debris was decanted. The protoplasts were washed with 0.7 M sorbitol, and finally sorbitol was replaced by 20% sucrose solution and centrifuged at 100 g for 10 min. Green protoplasts formed a band at the top of sucrose solution. They were transferred with a pipette to another tube. Protoplast culture medium was added and centrifuged at 100 g for 3 min. Such washings were repeated 3 times. The cleaned protoplasts floated; they were gently pipetted out with Pasteur pipettes and resuspended in the culture medium.

Glusulase, a commercially-available snail enzyme preparation, has also been used for the isolation of protoplasts from a number of leaf and root tissues (Pinto de Silva, 1969).

The cells that are damaged or lysed during isolation may release hydrolytic enzymes capable of damaging the healthy protoplasts. To counter this problem, addition of potassium dextran sulphate (0.5% w/v) to the enzyme solution has been recommended. Addition of antioxidants such as PVP-10 to the enzyme mixture has proved essential for the isolation of large numbers of viable protoplasts of recalcitrant plants such as deciduous tree species.

12.4. CULTURE OF PROTOPLASTS

Isolated protoplasts are cultured either in liquid medium or agar medium. The droplets of protoplasts suspension are placed in a liquid medium in petri dishes. It, in turn, is covered by another petri plate and sealed with paraffin. The culture dish is then maintained at 25–28°C.

Bergman's cell plating technique may be followed for culturing the protoplasts on an agar medium. A 2 ml aliquot of protoplasts is suspended in a liquid culture medium and mixed with an equal volume of the same medium containing 1% agar, the temperature of which should not exceed 45°C. On solidification of agar, the culture plates are sealed and maintained in an inverted position at 25–28°C. With this method, individual protoplasts or heterokaryons can be conveniently observed under microscope and the PE readily determined.

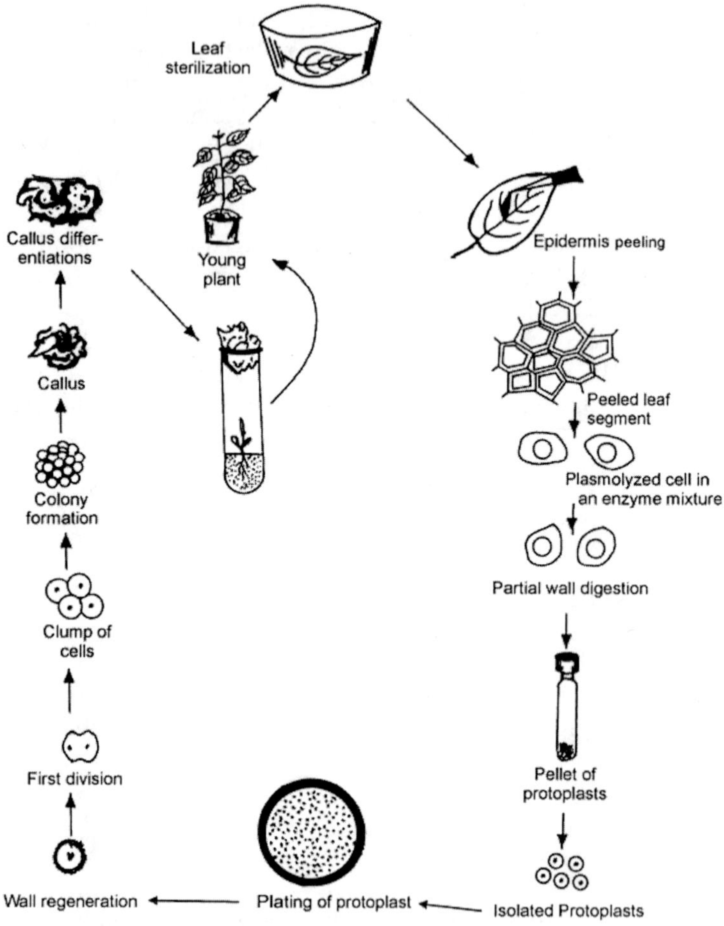

Fig. 39 Isolation, culture and regeneration of the plants from leaf protoplasts

Plating density: Protoplast concentration is critical for division. In liquid medium, the minimum concentration necessary is between 10^4–10^5 protoplasts per ml, whereas for agar plating it is between 10^3–10^4 protoplasts per ml.

Osmoticum: Isolated protoplasts require osmotic protection in the culture medium until they regenerate a strong wall. Osmolarity of the medium is generally adjusted with 500–600 mM/l mannitol or sorbitol, as in the enzyme solution. Seven to ten days after initial culture, by which time most of the viable protoplasts have regenerated a good wall and undergone a few divisions, the osmolarity of the medium is gradually reduced by periodic addition of a few drops of fresh medium lacking in the osmoticum or containing it at a fairly low level. In the presence of original high level of the osmoticum, the cells may stop dividing after some time. Macroscopic colonies are finally transferred to a fresh medium lacking the osmoticum.

Ionic substances (335 mM/l potassium chloride and 40 mM/l magnesium sulphate) improve the viability of protoplasts and yield cleaner preparations. Usually, enzyme solutions are supplemented with certain salts (5–100 mM/1 calcium chloride) along with non-ionic osmotic stabilizers. Cocking and Peberdy (1974) developed a CPW solution containing salts and a suitable osmoticum to provide stable and cleaner preparations. The CPW solution can be used during an enzyme incubation and washing of the protoplasts.

Nutritional requirements: The basic nutritional requirements for the culture of protoplasts are more or less similar to those of the cell culture. Mostly, the salts of MS and B5 media with some modifications have been found suitable. Generally, protoplast culture media contain 3–5% sucrose, but in tobacco protoplast culture, lower sugar (1.5%) content is required. Organic nitrogen in the form of CH is generally included in the medium as reduced nitrogen, and ammonium nitrate (20 mM/l) as reduced inorganic nitrogen. Antioxidants in the medium are either essential, or in some cases, they improve the response of the cultured protoplasts.

Physical treatments: Electroporation treatment of the isolated protoplasts has been shown to trigger early onset of cell divisions and bring about a significant increase of protoplast PE in herbaceous and woody species.

Nagata and Takebe (1971), for the first time, regenerated entire plantlets from mesophyll protoplasts of *Nicotiana tabacum*. The cell wall formation in cultured protoplasts usually takes place once the enzymes are removed. Soon after the formation of a wall around the protoplasts, the reconstituted cells show considerable increase in size, and first divisions generally occur within a week. Subsequent divisions give rise to small cell colonies. After 2–3 wk, macroscopic colonies are formed, which can be transferred to an osmotic-free medium to develop a callus. The callus may be induced to undergo organogenic differentiation, or whole plant regeneration (Fig. 37).

Plant regeneration may also occur via embryogenesis. Plant regeneration of the protoplasts via embryogenesis mostly occurs in species belonging to Poaceae, Apiaceae, Rutaceae and Fabaceae (forage legumes) among angiosperms, and Pinaceae of gymnosperms.

Since the first report of plant regeneration from protoplasts, the list of species exhibiting this potentiality has increased. Initially, the majority of taxa, which yielded protoplasts capable of regenerating callus or whole plants, belonged to Solanaceae. However, subsequent advances in the protoplast technology resulted in differentiation of shoots of the plants from isolated protoplasts of several other families including forage legumes, fibre or pulp crops, various tree species and even cereals, such as *Pennisetum*, rice and wheat.

12.5. APPLICATIONS

Besides being able to fuse with each other, higher plant protoplasts can also take up foreign DNA through their naked plasma membrane under specific chemical and physical treatments. Protoplasts also provide an experimental system for a wide range of biochemical and molecular studies ranging from investigations into the growth properties of individual cells to membrane transport. A direct way of transferring the cloned genes into the protoplasts by microinjection of DNA directly into the nucleus of tobacco mesophyll protoplasts was developed by Crossway *et al.* in 1986. Abdullah *et al.* (1986) demonstrated that the normal green rice plants can be regenerated efficiently and reproducibly from the protoplasts of rice via somatic embryogenesis. Shimamoto *et al.* (1989) produced transgenic fertile rice plants through regeneration from transformed protoplasts. Recovery of the fertile plants was made possible from the protoplasts of maize by the experiments of Prioli and Sondahl (1989).

12.6. PROTOPLAST FUSION AND SOMATIC HYBRIDIZATION

The technique of hybrid production through the fusion of protoplasts from different backgrounds is known as somatic hybridization.

1. **Spontaneous fusion:** Sometimes, during the enzyme treatment, it is observed that protoplasts from adjoining cells may fuse to form homokaryons (homokaryocytes). These plurinucleate cells sometimes contain 2–4 nuclei. Spontaneous fusion is of no significance.

2. **Mechanical fusion:** Isolated protoplasts are brought into intimate contact by mechanical means through micromanipulators and perfusion micropipettes. Protoplasts released from meiocytes (*Lilium* and *Trillium*) in enzyme solutions readily fuse by gentle tapping in a depression slide and some of the dinucleate and trinucleate cells reportedly complete the second division.

3. **Induced Fusion**

   a. **Treatment with sodium nitrate:** Power *et al.* (1970) was the first to induce protoplast fusion under the influence of sodium nitrate. They obtained intraspecific and interspecific hybrids between the protoplasts obtained from root tips of oat and maize seedlings. Carlson *et al.* (1972) used this method for producing the first somatic hybrid plant fusing protoplasts of *Nicotiana glauca* and *N. langsdorfii.* The transfer of the protoplasts in 0.25 M sodium nitrate solution and subsequent centrifugation promoted this process.

   b. **Calcium ions at high pH:** Keller and Melchers (1973) have studied the combined effect of calcium ions at high pH on fusion of tobacco protoplasts. This method involves spinning the protoplasts in a fusion-inducing solution (0.05 M calcium chloride in 0.4 M mannitol at pH 10.5) for 3 min at 50 g, and then keeping the tubes in a water bath (37°C) for 40–50 min. Aggregation of the protoplasts, generally takes place at once and fusion occurs in 10 min.

   c. **PEG treatment:** The PEG has been used as a fusogen in a number of plant species. About 0.6 ml of PEG solution (dissolve 1 g of PEG in 2 ml of 0.1 M glucose, 10 mM calcium chloride and 0.7 mM potassium phosphate) is added in drops to a pellet of protoplasts in the tube. The tubes are incubated at room temperature for 40 min. Occasional rocking of the tubes helps to bring the protoplasts in contact. This is followed by elution of PEG by the addition of 0.5–1 ml of protoplast culture medium in the tube after every 10 min. Preparations are now washed free of fusogen by centrifugation and the protoplasts resuspended in the culture medium. Compared to low-frequency fusion by sodium nitrate, the PEG induced fusion gives better results.

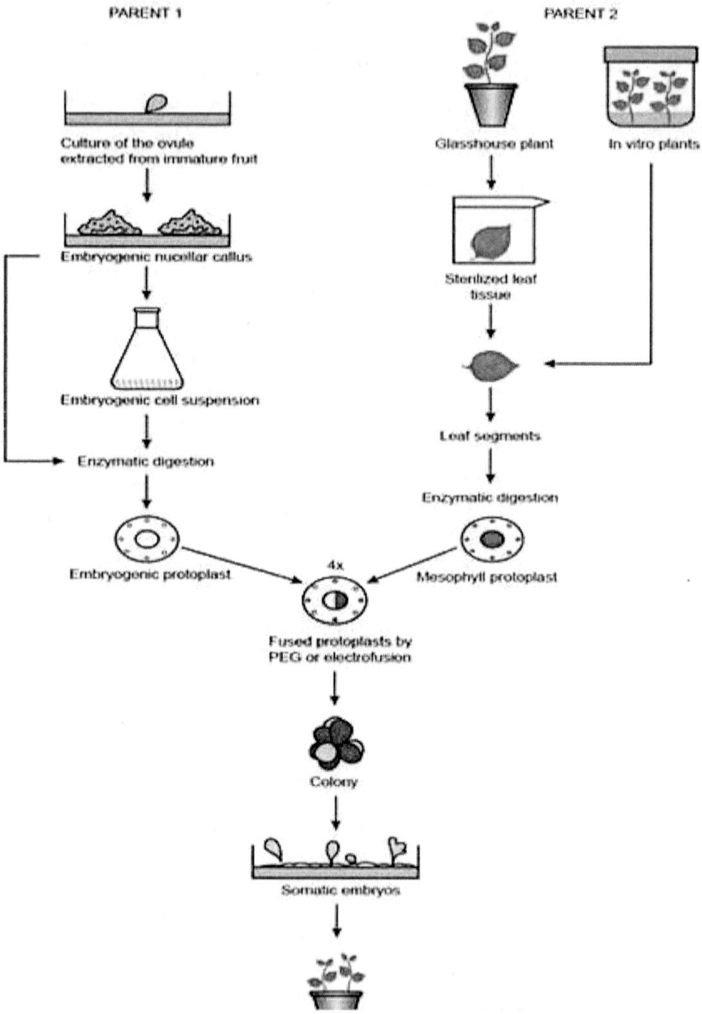

Fig. 40 Development of plantlet through somatic hybridization by PEG treatment

d. **Electrofusion:** Senda *et al.* (1979) first attempted fusion by positioning two microelectrodes with the help of micromanipulator at the ends of adjoining *Rauvolfia* protoplasts. Success in inducing fusion was achieved with brief 5–12 amp DC pulses restricted to single protoplast pairs only. Subsequently, Zimmerman and Scheurich (1981) demonstrated that the batches of protoplasts could be fused by electric fields by devising a protocol, which is now widely used. This protocol involves two-step process. First, the protoplasts are introduced into a small fusion chamber containing parallel wires or plates, which serve as electrodes. Second, a low-voltage and rapidly-oscillating AC field is applied, which causes protoplasts to become aligned into chains of cells (pearl chains) between the electrodes. This creates complete cell-to-cell contact within a few minutes. Once alignment is complete, the fusion is induced by application of a brief spell of high-voltage DC pulses (0.125–1 kv cm^{-1}). A high voltage DC pulse induces a reversible breakdown of the plasma membrane at the site of cell contact, leading to fusion and consequent membrane reorganization. The entire process, starting from the introduction of the protoplasts inside the chamber and their transfer to culture media, can be completed in 5 min or less. Shoot or somatic hybrid plant regeneration after electrofusion of the protoplasts has been reported in combinations: *Nicotiana tabacum* (+) *N. tabacum*, *N. plumbaginifolia* (+) *N. tabacum*, *N. glauca* (+) *N. langsdorfii* and *Solanum tuberosum* (+) *S. phureja*.

Induced fusion is influenced by an alkaline pH and high calcium concentration. The process of fusion is considerably enhanced if the protoplasts suspension is pretreated with a membrane-activating agent such as lysozyme.

Fusion product of the protoplasts of same species is known as homokaryon while the fusion product of protoplasts of different species is known as heterokaryon.

The first report of successful production of a mature interspecific hybrid plant by fusion of protoplasts is by Carlson *et al.* (1972) between *Nicotiana glauca* + *N. langsdorfii*. Power *et al.* (1976) obtained interspecific somatic hybrids of *Petunia hybrida* + *P. parodii*.

An outstanding example of intergeneric fusion of protoplasts is the work of Melchers *et al.* (1978). They reported regeneration of the somatic hybrid plants of potato and tomato. Mesophyll protoplasts of *Lycopersicon esculentum* and protoplasts obtained from liquid callus culture of *Solanum tuberosum* were fused in the presence of PEG with a high calcium ion concentration. The fused protoplasts divided and developed into callus masses, some of which showed regeneration of the green shoots. The shoots were either transferred to soil or grafted on to a tomato stock.

12.6.1. Selection of Somatic Hybrids

Following the fusion treatment, the protoplast population consists of parental types, homokaryotic fusion products of parental cells, heterokaryotic fusion products or hybrids and a variety of other nuclear cytoplasmic combinations. Despite efforts to increase the efficiency of protoplast fusion, usually not more than 1–10% of the protoplasts in a treated population have actually undergone fusion. The proportion of viable heterospecific binucleate A + B type fusion products may be even lower. In the absence of systems that would fuse different protoplasts specifically to produce heterokaryocytes at a very high rate, there is an obvious need to select the products of fusion amongst the unfused and homofused parental cells. Identification and recovery of the protoplast fusion products have been based on: 1. Observation of visual characters, 2. Hybrid cells display genetic complementation for recessive mutations, and 3. Physiological complementation for *in vitro* growth requirements. In complementation, fusion of two protoplasts, each carrying a different recessive marker, will generate a fusion product, which is functionally restored since each parent contributes a functional allele that corrects the respective deficiency of the other parent.

- **Auxotroph complementation:** This method of selection has been applied to higher plants on a limited scale. The limitation is due to the paucity of higher plant auxotrophs. There is a report by Glimelius *et al.* (1978) where two parents were NR deficient mutants of *N. tabacum* and could not be grown with nitrate as a sole nitrogen source while the hybrids could regenerate shoots in the nitrate medium. The lack of NR activity causes an absolute requirement for reduced nitrogen and is caused by a deficiency either in the NR apoenzyme (nia-type of mutant) or in the molybdenum cofactor (cnx-type mutant).
- **Drug sensitivity:** Power *et al.* (1976) utilized the differential sensitivity of the protoplasts isolated from *Petunia parodii* and

P. hybrida to the drug actinomycin D. In MS medium, the mesophyll protoplasts of *Petunia hybrida* develop up to a macroscopic callus stage and those of *P. parodii* divide to form only small cell colonies. The addition of actinomycin D to the culture medium apparently has a little effect on the regeneration potential of *parodii* protoplasts, but those of *P. hybrida* fail to divide. Heterokaryons, however, are able to grow despite the presence of the drug and ultimately differentiate into somatic hybrid plants (Fig. 41). A similar procedure was adopted in the selection of somatic hybrids between *Nicotiana sylvestris* and *N. knightiana* (Maliga *et al.,* 1977).

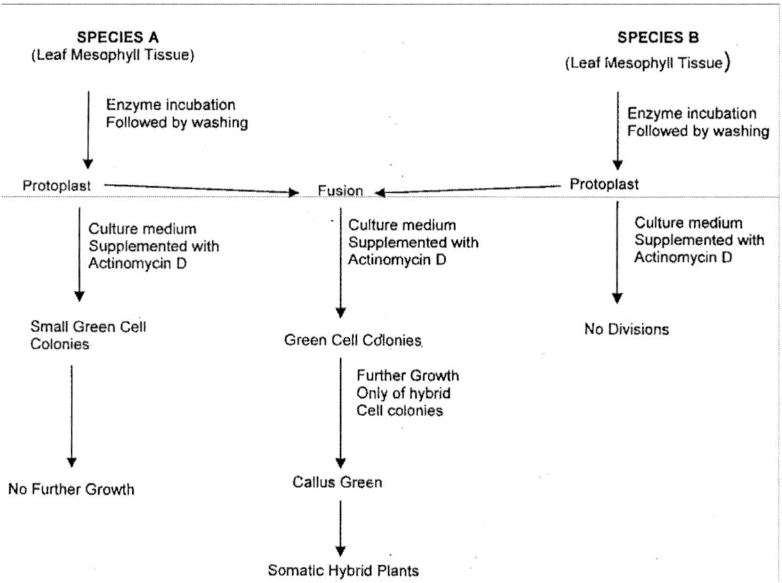

Fig. 41 Scheme illustrating the selection of somatic hybrids utilizing the differential sensitivity of the mesophyll protoplasts to drug actinomycin (e.g., interspecific somatic hybrids between *Petunia parodii* and *P. hybrida*)

- **Use of visual characteristics:** The most efficient, but the most tedious method to select the products of protoplast fusion, is to visually identify the hybrid cells and mechanically isolate the individual cells. The approaches used by various workers are given below:

- **Use of morphologically-distinct cells:** The fusion of mesophyll protoplasts containing green chloroplasts with colourless cell-culture protoplasts containing distinct starch granules due to growth on sucrose-supplemented medium. Fused products can be seen immediately after PEG treatment with one half containing chloroplast and the other half with distinct starch granules (Fig. 42)

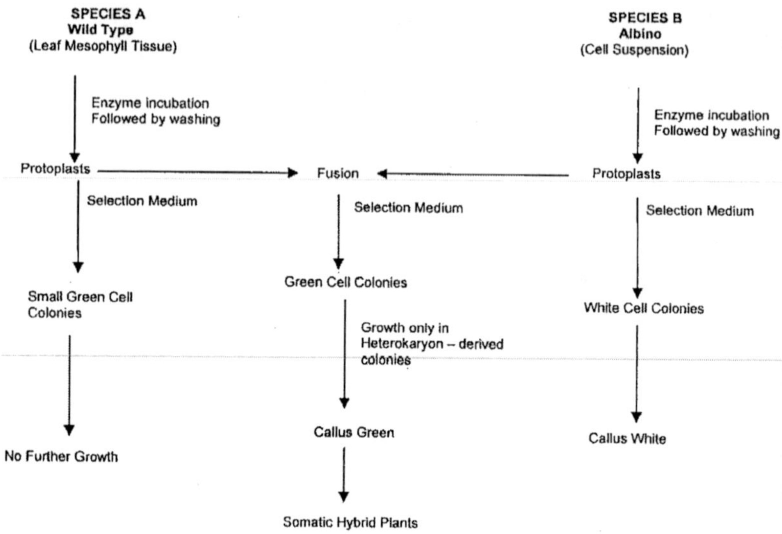

Fig. 42 Visual selection procedure coupled with differential growth of parental protoplasts to media constituents enabling only hybrid callus to regenerate plants

- **Use of petal protoplasts** with leaf protoplasts and petal + cell culture protoplasts can be readily distinguished. The usually vacuolar petal pigment is originally separated within the fused cell, but eventually becomes evenly-distributed throughout the fused cell (Fig. 43).

- **Use of non-allelic albino mutants for complementation selection:** This selection system was developed by Melchers and Labib (1974). They fused haploid chlorophyll-deficient and light-sensitive protoplasts of *Nicotiana tabacum,* and cultured them under high intensities of light. After 2 mo, green colonies were observed in the culture dishes as a consequence of complementation between the two albino mutants. On further culturing, these green colonies regenerated somatic hybrid plant. Non-allelic albino mutants have been used successfully to produce intraspecific (*Datura innoxia* + *D. innoxia*) and interspecific (*Petunia parodii* + *P. hybrida,* Fig. 41) somatic hybrids.

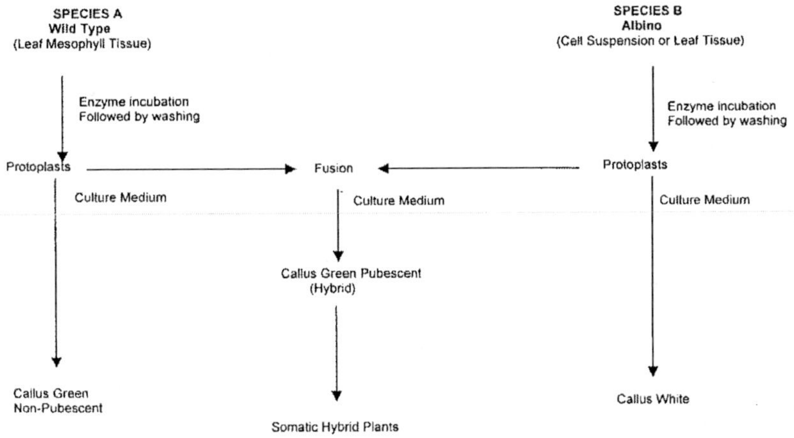

Fig. 43 Selection scheme used in the intergeneric somatic hybridization of *Atropa belladonna* (wild type) with *Datura innoxia* (albino)

Chlorophyll-deficient mutants have also been used to raise the somatic hybrids of *N. tabacum* + *N. sylvestris*.

12.7. CYBRIDS

Sexual hybridization is a precise mixture of parental nuclear genes, but the cytoplasm is of maternal parent only. On the other hand, in somatic hybrids, the cytoplasm is derived from both the parents. However, somatic hybrids can be obtained where nucleus is derived from one parent and cytoplasm is derived from both, thus producing cytoplasmic hybrids, also called as cybrids. Early segregation of the nuclei in a fused product can be stimulated and directed so that one protoplast contributes the cytoplasm while the other contributes the nucleus alone or both nucleus and cytoplasm (Fig. 45). There are different ways of inactivating the nucleus of one protoplast. Thus, there will be a fusion between protoplasts containing the full complement of nucleus, mitochondria and chloroplasts with cytoplasmic component of second parent. The various approaches to achieve this type of fusion are:

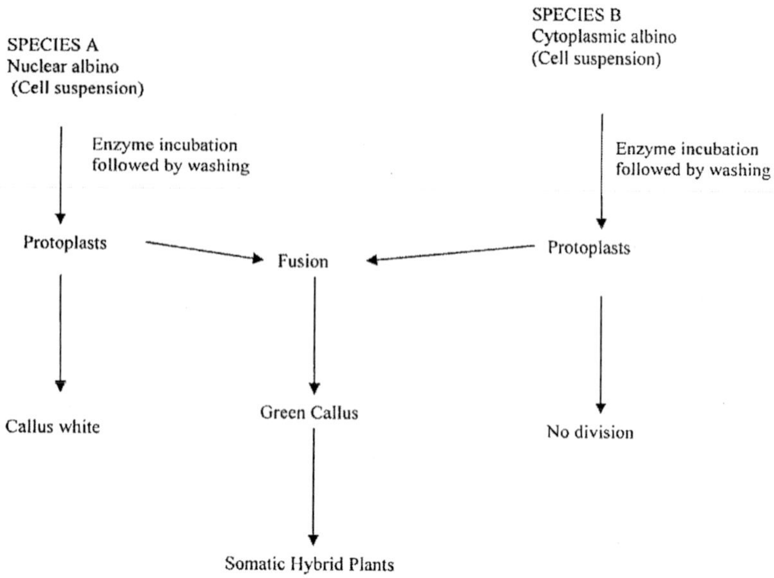

Fig. 44 Use of non-allelic albino mutants for complementation and selection of somatic hybrids (e.g., *Petunia parodii* (+) *P. hybrida*)

- **Application of lethal dosage of X-rays or gamma irradiation to one parental protoplast population:** This treatment renders the protoplasts inactive and non-dividing, but they serve as an efficient donor of cytoplasmic genophore when fused with recipient protoplasts. *Nicotiana* protoplasts can be inactivated by 5 kr dose of X-rays. Other protoplasts may require different doses.

- **Treatment with iodoacetate to metabolically inactivate the protoplasts:** Pretreatment with iodoacetate will cause the degeneration of non-fused and autofused protoplasts while fusion of iodoacetate-pretreated protoplasts with non-treated protoplasts will cause metabolic complementation and result in viable hybrids. In an experiment, iodoacetate-treated *Nicotiana plumbaginifolia* cell suspension was fused with X-irradiated *N. tabacum* mesophyll protoplasts. All the regenerated cybrid plants had *N. plumbaginifolia* morphology, but most of them contained *N. tabacum* chloroplasts. The iodoacetate treatment does not impair the nucleus of the treated protoplasts. Thus, the latter can complement the X-irradiated protoplast.

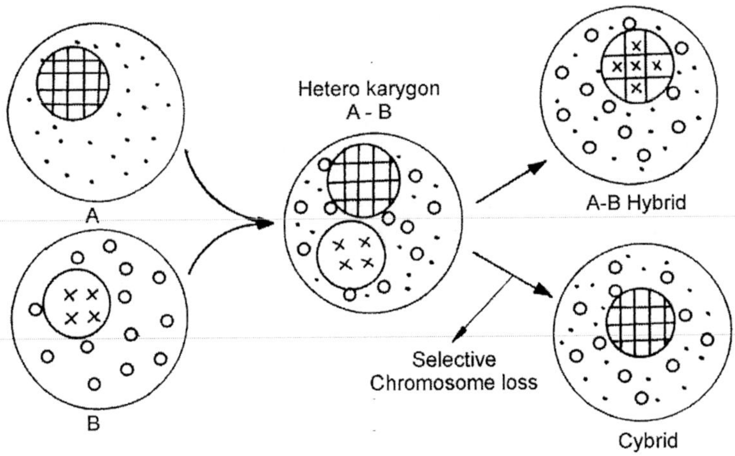

Fig. 45 Mechanism of fusion of the protoplasts

- **Fusion of normal protoplast with enucleated protoplasts:** Medgyesy *et al.* (1980) transferred streptomycin resistance (controlled by chloroplast DNA) from *Nicotiana tabacum* to *N. sylvestris* by fusing iodoacetate-treated, non-dividing protoplast of streptomycin resistant *N. tabacum* with normal protoplasts of streptomycin-sensitive *N. sylvestris*. Enucleated protoplasts can be obtained by high-speed centrifugation (20,000–40,000x g) for 45–90 min in an isosomotic density gradient with 5–50% percoll.
- Fusion of abnormal protoplast with another in which nuclear division is suppressed

12.8. APPLICATIONS

Pirrie and Power (1986) succeeded in producing fertile, interspecific, gametosomatic, triploid hybrids of tobacco by fusing the protoplasts of leaf (diploid) and the protoplasts of the pollen tetrad (haploid). This was followed by the production of somatic hybrids between *Lycopersicon esculentum* and *L. peruvianum* by Kinsara *et al.* (1986). The plantlets from the somatic hybrid cells of *Oryza sativa* and *Echinochloa oryzicola* were regenerated by Terad *et al.* in 1987. Later, tetraploid and hexaploid somatic hybrids from protoplast fusions between *Solanum brevidens* (diploid, non-tuber bearing species), and diploid and tetraploid *Solanum tuberosum* were produced by Ethenfelt and Helgeson (1987).

- Production of novel interspecific and intergeneric crosses between the plants that are difficult or impossible to hybridize conventionally. It overcomes incompatibility barriers. An outstanding example of intergeneric fusion of protoplasts is the work of Melchers *et al.* (1978). They reported regeneration of somatic hybrid plants of potato and tomato. Efficient cell fusion can be achieved between sexually-compatible and incompatible parents involving interspecific and intergeneric combination (Table 11).
- In plant breeding programmes, sexual crossings at interspecific and intergeneric levels often fail to produce hybrids due to incompatibility barriers. The bottlenecks in sexual hybridization may, therefore, be overcome by somatic cell fusion. In some cases, somatic hybrids between two incompatible plants have also found application in industry and agriculture.

Schnieder (1978) obtained amphidiploid *Datura innoxia* (+) *D. discolor* and *D. innoxia* (+) *D. stramonium*, by fusing their diploid mesophyll protoplasts. These hybrids did not exist in nature as conventional breeding procedures proved unsuccessful. Somatically-produced amphidiploids of these combinations of *Datura* species are propagated for industrial uses as they demonstrate heterosis and higher (20–25%) scopolamine content than in the parental forms.

Nicotiana repanda, *N. nesophila* and *N. stocksonii* are resistant to a number of diseases, but are not sexually crossable with tobacco (*N. tabacum*). However, fertile hybrids have been reported in combination such as *N. tabacum* + *N. nesophila* and *N. tabacum* + *N. stocksonii* by protoplast fusion. Somatic hybridization of dihaploid and tetraploid potato protoplasts with isolated protoplasts of *Solanum brevidens*, *S. phureja* and *S. pennellii* resulted in the synthesis of fertile, partially-amphidiploid plants possessing important agricultural traits e.g., resistance to potato leaf virus, potato virus Y and *Erwinia* soft rot. Interspecific somatic hybridization involving species that are sexually incompatible with brinjal (*Solanum melongena*) has also resulted in the production of amphidiploids with traits resistant to *Verticillum* wilt. Somatic hybridization, between *Brassica napus* and *B. nigra* cultivar, possessing the gene for resistance to *Phoma lingam*, yielded amphidiploid plants, carrying this gene for resistance. These hybrids possess all the three *Brassica* genomes (A, B and C) and are now incorporated in breeding programmes.

- **Cytoplasmic male sterility:** Several agriculturally-useful traits are cytoplasmically encoded, including some type of male sterility, and certain antibiotic and herbicide resistance factors. Pelletier *et al.* (1988) reported *Brassica–Raphanus* cybrids that contain the nucleus of *B. napus*, chloroplasts of atrazine-(herbicide) resistant *B. campestris* and mitochondria that confer male sterility from *Raphanus sativus*. Rambaud *et al.* (1997) reported that CMS chicory (*Cichorium intybus*) cybrids have been obtained by the fusion between chicory and CMS sunflower protoplasts. Using protoplast fusion, CMS has been transferred from *Nicotiana techne* to *N. tabacum* and *Petunia hybrida* to *P. axillaris*. Zelcer *et al.* (1978) transferred CMS by protoplast fusion from *Nicotiana tabacum* to *N. sylvestris*.
- **Herbicide resistance:** Resistance of the plants to herbicide atrazine has been transferred from *Brassica campestris* to *B. napus* via fractionated protoplast fusions.

Table 11: Intergeneric hybrids produced through protoplast fusion

Arabidopsis thaliana + Brassica campestris
Barbarea vulgaris + Brassica napus
Brassica campestris + Barbarea vulgaris
Brassica juncea + Diplotaxis muralis
Brassica juncea + Eruca sativa
Brassica juncea + Trachystoma baillii
Brassica oleracea + Moricandia arvensis
Brassica napus + Eruca sativa
Brassica napus + Thlaspi perfoliatum
Citrus aurantifolia + Feronella lucida
Citrus aurantifolia + Swinglea glutinosa
Citrus reticulata + Citropus gabunensis
Citrus sinensis + Atalantia ceylanica
Citrus sinensis + Murraya paniculata
Citrus sinensis + Poncirus trifoliata
Citrus sinensis + Severinia disticha

Datura innoxia + Atropa belladonna
Daucus carota + Aegipodium podagrina
Daucus carota + Petroselenium hortense
Oryza sativa + Echinochloa oryzicola
Sinapsis turgida + Brassica oleracea
Sinapsis turgida + Brassica nigra
Solanum tuberosum + Lycopersicon esculentum
Solanum tuberosum + Lycopersicon pimpinellifolium

Table 12: Genetic traits transferred via protoplast fusion

Somatic hybrids	Traits
Resistance	
Nicotiana tabacum + N. nesophila	TMV
N. tabacum + N. nesophila	Tobacco hornworm
Solanum tuberosum + S. chacoense	Potato virus X
S. tuberosum + S. brevidens	PLRV Late blight and PLRV Potato virus Y
S. circaeifolium + S. tuberosum	*Phytophthora*
S. melongena + S. saintwongsei	*Pseudomonas solanacearum*
S. tuberosum + S. comersonii	Frost tolerance
Brassica napus + B. nigra	*Phoma lingam*
Brassica oleracea + Sinapsis alba	*Alternaria brassicae*
Brassica napus + Sinapsis alba	*Alternaria brassicae*
Citrus sinensis + Poncirus trifoliata	*Phytophthora*
Lycopersicon esculentum + L. peruvianum	TMV, spotted wilt virus, cold tolerance
Solanum lycopersicoides + L. esculentum	Cold tolerance
Solanum ochranthum + Lycopersicon esculentum	Tomato diseases and insect pest
Brassica oleracea + B. napus	Black rot (*Xanthomonas campestris*)
B. oleracea ssp. *capitata + B. oleracea* (Ogura CMS line)	Cold tolerance
B. oleracea var. *botrytis + Sinapsis alba*	*Alternaria brassicola*

Somatic hybrids	**Traits**
Raphanus sativus + *B. oleracea* var. *botrytis*	Club rot disease
Citrullus lanatus + *Cucumis melo*	Club rot resistance
Hordeum vulgare + *Daucus carota*	Frost and salt tolerance
Solanum melongena + *S. sisymbrifolium*	Nematode
S. tuberosum + *S. bulbocastanum*	Nematode, Root knot nematode
Raphanus sativus + *Brassica napus*	Beet cyst nematode
Sinapsis alba + *Raphanus sativus* + *B. napus*	Beet cyst nematode
Quality characters	
Nicotiana rustica + *N. tabacum*	High nicotine content
Brassica napus + *Eruca sativa*	Low erucic acid
Datura innoxia (+) *D. discolor*	Higher scopolamine content
D. innoxia (+) *D. stramonium*	Higher scopolamine content
Agronomic characters (transferred via cybrid formation)	
Nicotiana tabacum + *N. sylvestris*	Streptomycin resistance
Solanum nigrum + *S. tuberosum*	Atrazine resistance
Brassica nigra + *B. napus*	Hygromycin resistance
Nicotiana tabacum + *N. sylvestris*	CMS
Brassica campestris + *B. napus*	CMS
B. napus + *B. campestris* + *Raphanus sativus*	CMS and atrazine resistance
Lycopersicon esculentum + *Solanum acaule*	CMS
Brassica napus + *B. tournefortii*	CMS

13 SOMACLONAL VARIATIONS

13.1. INTRODUCTION

Recent investigations have revealed that the cell or tissue cultures undergo frequent genetic changes (polyploidy, aneuploidy, chromosomal breakage, deletion, translocation, gene amplifications and mutations) and that these are also expressed at biochemical or molecular levels. Plant tissue cultures, therefore, provide an increased genetic variability; relatively rapidly and without applying a sophisticated technology.

Usually, tissue culture is considered as a method of developing clones of a particular individual plant or genotype, and is understood that all the regenerants from a tissue culture system will be replicas of the donor plant. However, very often, deviants or variants from the source plant have been noticed among the regenerants, which then were regarded as 'artifacts of tissue culture'. Some considered them as 'epigenetic variants' probably caused by exposure to phytohormones. Skirvin and Janick (1976) called them as 'calliclone variants' and emphasized their probable role in crop improvement. Later, Larkin and Scowcraft (1981) suggested the term 'somaclonal variation' for such variants among the tissue culture regenerants.

13.2. ORIGIN OF SOMACLONAL VARIATION

An understanding of the causes of somaclonal variation is important either for enhancing or suppressing the production of such variants. Though the actual mechanism is not clearly understood, the possible causes are:

- Some of the variations could be revelation of the inherent cellular heterogeneity of the explants.

- Culture conditions may bring about new genetic changes.

13.2.1. Pre-existing Variability

Plant development, in general, involves change in nuclear DNA such as change in chromosome number, structure (Bennici and D'Amato, 1990). Cells of the plant apical meristems such as root-tips and shoot-tips are uniformly diploid in their genome due to DNA synthesis immediately followed by karyokinesis and cytokinesis (normal cell cycle). However, the derivatives of these meristematic cells do not divide by normal mitosis, but may undergo DNA duplication and endoreduplication. The varying degree of endoreduplication results in somatic cells with 4C, 8C or higher DNA content or may result in polysomaty. Usually, these genetic changes are not noticed as these cells do not divide. However, under culture conditions, these cells may divide, and undergo redifferentiation and express this change in their genome content as an inheritable character within the whole plant. Another type of pre-existing chromosomal variability, which is rarely observed in hybrid plants, is aneusomaty. In such plants, the apical meristems and, consequently, the mature tissues comprise a mosaic of cells with varying number of aneuploid chromosome numbers. This condition is transferred or enhanced in callus cultures derived from such tissues.

13.2.2. Growth Pattern and Regeneration Mode

In vitro growth may occur from meristem cultures, which may form callus (undifferentiated mass of cells) or direct shoot formation. Callus is further differentiated into organized structures by organogenesis or somatic embryogenesis. The departure from organized growth is a key element in somaclonal variation. In general, longer the duration of callus and cell suspension in culture phase, the greater the chances of generating somaclonal variation. These somaclonal variations can also occur in embryogenic cultures if they are kept for a long time in cultures, depending upon the plant species.

13.2.3. Length of Culture and the Subculture Frequency

The age of callus and cell suspension in culture has a marked effect on the frequency of somaclonal variation, which increases with the increase in the duration of *in vitro* cultures. The reduction or even total loss of regeneration ability is a general phenomenon observed in many long-term callus or cell culture lines. The level of DNA polymorphism has been found to be increased with length of time in culture. Although the chloroplast genome is, generally, considered to be more highly-conserved and stable than nuclear and mitochondrial genomes, the prolonged culture resulted even in the deletion of parts of the chloroplast genome. These deletions were associated with the changes in plastid morphology. The age affects the frequency of mutation, which is primarily due to sequential accumulation of mutations over time than an increased mutation rate in old cultures. The frequent subcultures of the callus and cell suspension favours the stability in ploidy and chromosomal constitution compared to extended subculture intervals.

13.2.4. Mechanisms Leading to Genetic Variations

- **Gross karyotypic changes:** Polyploidy and aneuploidy observed in some cultures such as tobacco, *Pelargonium* etc. might be responsible for the production of variants. Karyotype changes: variant plants with altered chromosome number have been reported by several workers. Polyploidy is the most frequently-observed chromosome abnormality e.g., aneuploidy in oats (McCoy, 1982), potato (Sreeramulu *et al.,* 1985) and barley (Kole and Chawla, 1993). The change in chromosome number in a variant plant is commonly associated with reduced fertility and with altered genetic ratios in the progeny of self-fertilized plants.

- **Cryptic chromosomal structural changes:** This is considered as a major mechanism to generate somaclonal variants. Chromosome breakage, reunion, centric as well as acentric fragments, deletions, translocations and inversions have been observed in several cultures such as barley. Each of them might have contributed to the variations; some of the variants could also be the result of position effects. It is also possible that some segmental haploid states or hemizygosities may occur due to deletions. This has been called as 'culture-induced hemizygosity' by Siminovich (1976) and may lead to sudden expression of recessive characters. More cryptic chromosome rearrangements may be responsible for genetic variation. Translocations have been reported in potato, ryegrass, oats etc. Published work also suggests that chromosome deletions, duplications, inversions, and other minor reciprocal and non-reciprocal rearrangements occur among regenerated plants e.g., ryegrass, barley. Chromosome irregularities such as breaks, acentric and centric fragments, ring chromosomes and micronuclei were observed in garlic somaclones. Cytological abnormalities such as multilobed nuclei, multinucleate cells, abnormal anaphase and mixoploidy were observed in SC3 generation of barley.

- **Somatic crossing over and sister-chromatid exchange:** Somatic crossover leading to sister-chromatid exchange, though originally rare, can be enhanced by certain chemicals such as BuDR, mitomycin and physical variants such as X-rays, temperature etc. Some of the chemicals in the medium might probably contribute to an increase of such chromatid exchanges. When asymmetric exchanges occur, new gene combinations are produced, which may ultimately lead to the development of new characters or suppression or change in the expression of existing characters due to position effects. The MCO can also account for some of the variation detected in regenerated plants. This can include both symmetric and asymmetric recombination. The MCO may also account for the recovery of homozygous recessive single gene mutations in some regenerated plants. Somatic cell sister-chromosome exchange, if asymmetric, can also lead to deletion and duplication of genetic material, and hence, variation.

- **Transposable elements:** Transposable genetic elements are DNA portions that can change their position in the genome. Their excision from a place and reinsertion at another place may lead to chromosomal breakages as well as disruption of the adjacent gene functions. Tissue culture media might be conducive for such transposon activation leading to variant production. Such high mutability and adaptiveness in somatic tissues have been referred to as somatic Darwinism (Weill and Reynaud, 1980).

- **Gene amplification and diminution:** A cell can increase or decrease the quantity of a specific gene product by differential gene amplification or diminution by suitable regulation mechanisms. As phytohormones have genetic regulatory effect, such gene amplifications or diminutions may occur under culture conditions leading to production of variants.

Soybean cultures adapted to prolonged growth on maltose have lost one-third of their rRNA genes. This diminution can be reversed by readaptation of the cells to sucrose media; provided the maltose adaptation was not prolonged.

In maize, serial selection of sensitive cell lines to T-toxin of *Drechslera maydis* resulted in cell lines resistant to high-toxin levels. Similarly, Nabors *et al.* (1980) used 11 successive selection cycles to obtain high-level resistance of salinity in tobacco cultures. Such selections are thought to be possible because of gene amplification in the culture.

- **Single-gene mutation:** Variants may also arise as a result of more subtle changes due to single-gene mutations in cultures, which have cells that are apparently showing no karyotype changes. Recessive mutations are not detected in RO Plants (plants regenerated *in vitro* from any cell or tissue), but express in RI progeny (the progeny obtained after selfing of RO plants). The RI progeny segregates in Mendelian 3:1 ratio for the trait of interest. This further confirms the mutant nature of the variant. Somaclonal variants for single recessive gene mutations have been reported in respect of maize, *Nicotiana sylvestris*, rice and wheat.
- **Cytoplasmic genome changes**: The most detailed work on this has been done on maize. Many studies have implicated mitochondria as the determinants of male sterility and *Drechsleria maydis* T-toxin sensitivity in maize. Gengenbach *et al.* (1977) and Brettell *et al.* (1979) regenerated plants that were resistant to fungal T-toxin. The original lines used had all the Texas male sterile cytoplasm, which results in plants being normally highly susceptible to this disease. However, when selection pressure was applied by subculturing callus cultures in the presence of toxin, some regenerants were found to be not only resistant to the toxin, but also fertile.

13.2.5. Analysis of Somaclonal Variants

Most useful somaclones are those, which carry almost all of the good parental characters as well as incorporate certain desirable characters within it; the characters that were lacking in its parents. It becomes extremely important to select variants as early as possible, with minimal exposure of the cells to tissue culture environment. Gross abnormalities may appear with prolonged culture. The variants are generally assessed at the phenotypic level, and in over 50% cases, it is based on R_0 plants.

However, this approach of screening R_0 plants would the screening of only homozygous or dominant traits. The recessive mutations in heterozygous regenerants can be recognized only in the segregating R_1 and R_2 progenies. It is, therefore, important that the variants should be assessed in the sexual progenies of the *in vitro* regenerated plants so that their heritability is established.

The degree of variation of a plant can be determined by estimating the standard deviation for a particular quantitative trait. It is usually determined as the percentage of the plants showing aberrations for one or more defined characteristics such as plant height, time of flowering, fertility, flower and fruit colour. The effect of environment on the phenotype of plant can also be detected using biochemical characterization mostly involving protein electrophoresis. The above-mentioned methods can be very well used for the assessment of phenotypic variations, but the variation or change at genome level cannot be monitored. In order to detect the variation at DNA level, use of certain molecular markers is encouraged. The RFLP appears to be a better technique as it helps in identifying slight changes and also in studying plants grown in different environments.

13.2.6. Factors Influencing Somaclonal Variation

- **Chromosomal status of the explants:** The chromosomal composition (diploid, anueploid, polyploidy) of the explant gets multiplied in the culture and regenerants may originate from different ploidy levels. Greater the heterogeneity in the chromosomal status of the explant, greater will be the chromosomal variation in the regenerants.
- **Age of culture:** Chromosomal and genetic abnormalities were found to increase with increase in the age of the culture. Hence, older the culture or more the number of subcultures, greater will be the variation in the regenerants.
- **Phytohormones and other chemicals in the medium:** As the concentration of phytohormones such as 2, 4-D increases, the frequency of the chromosomal abnormalities in the culture are also found to increase, thus leading to the production of more variants.

13.2.7. Significance of Somaclonal Variations

Somaclonal variation may provide a source of increasing variability or genetic diversity needed for the crop improvement, especially in the self-pollinated and asexually-propagated systems. Somaclonal variation, cellular selection and early rapid screening of regenerants collectively provide a powerful option for plant improvement. The following are some classical examples of agronomically-important somaclones.

13.3. ISOLATION OF VARIANTS

- **Without selection pressure:** Unorganized callus and cells, grown in cultures for various periods on a medium that contains no selective agent (toxic or inhibitory substance), are induced to differentiate whole plants. The regenerated plants are ultimately transferred to the field and screened for variation. Somaclonal variants of various species have been obtained using this approach.
    - **Sugarcane:** The potential value of variant plants isolated from cell and tissue cultures was first recognized in respect of sugarcane. Krishnamurthi and his co-workers in Fiji isolated subclones of established cultivars of sugarcane through tissue culture. These cultivars showed resistance to eyespot disease (caused by *Helminthosporium sacchari*), Fiji disease (caused by an aphid-transmitted virus) and downy mildew (caused by *Sclerospora sacchari*). Similarly, Liu (1981) identified a calliclone of sugarcane superior to the best local Taiwanese variety (F-160) in cane yield, sugar yield and resistance to smut disease caused by *Ustilago scitaminea*.
    - **Potato:** Shepard *et al.* (1980) regenerated a large number of plants from mesophyll protoplasts of 'Russet Burbank' and recorded considerable variation in protoclone population. Some of them were resistant to early blight (*Alternaria solani*) or late blight (*Phytophthora infeastans*). Protoclones resistant to virus Y and leaf-roll virus (PVY) have also been observed in the field (Thomson *et al.,* 1986).
    - **Tomato:** Evans *et al.* (1984) isolated tomato somaclones with variant phenotypes such as recessive mutations for male sterility, resistance to *Fusarium oxysporum*, jointless pedicel, tangerine virescent leaf, flower and fruit colour.

Some of these variations in phenotypes were due to changes in chromosome number. At the DNA Plant Technology Co., USA, Evans *et al.* (1987) developed somaclonal variants of tomato-bearing fruits, which had enhanced taste, a better texture, colour and high (20%) dry matter.

- **Banana:** Israeli *et al.* (1991) raised banana plants using *in vitro* techniques exhibiting a range of morphological variations – dwarfism, abnormal leaves, pseudostem pigmentation, persistence of flowers and split fingers. All the commercial banana cultivars are susceptible to *Fusarium* wilt disease, and traditional breeding methods to select resistant lines have not been successful. Somaclones of Cavendish banana, on the contrary, showed resistance to the pathogen.
- **Cereals and grasses:** The callus-derived plants of cereals and grasses can be a rich source of somaclones. Plants of these clones vary in height, size, shape of leaves, length of awns, fertility of tillers or spikes, or colour of seeds. Variants selected for their commercial potential are male sterility, gliadin protein regulation, frost resistance in wheat, increased yield in rice, and earliness in maize.
- **Lucerne:** Protoclones of lucerne (*Medicago sativa*) regenerated from isolated mesophyll protoplasts reportedly showed variation for yield and disease resistance to *Fusarium solani, F. roseum* and *Rhizoctonia solani*. Some of these clones gave 60% higher yield than the parent lines.

- **With selection pressure:** In this method, the variant cell lines are screened from the cultures by their ability to survive in the presence of a substance in the medium that may be toxic/inhibitory or under conditions of environmental stress. Variants may be obtained by direct selection, indirect selection or by individually-testing single cell-derived colonies.
  - **Disease resistance:** Carlson (1973) reported that the plants regenerated from EMS-treated tobacco protoplasts, selected for resistance to MSO, showed enhanced tolerance to infection by *Pseudomonas tabaci*. The resistance was inherited as a single semi-dominant trait. Selection for resistance to *Helminthosporium* T-toxin (host-specific toxin) in maize cultures was reported along similar lines by Gengenbach *et al.* (1977). In their study,

the embryogenic maize calli were exposed to the toxin in the cultures, and in the process, they obtained resistant and fertile plants. The calli were derived from plants inheriting toxin-susceptible and male sterile traits maternally. Since cytoplasmic conversion to male fertility and T-toxin resistance did not transmit sexually in selfed progeny of T-cytoplasm maize, this example represents a typical case of somaclonal variation.

- **Salt tolerance:** Plant tissue culture technique has been successfully used to obtain salt-tolerant cell lines or variants in several plant species *viz*., tobacco, alfalfa, rice, maize, *Brassica juncea*, *Solanum nigrum*, sorghum etc. Mandal *et al.* (1999) developed a salt-tolerant somaclone BTS24 from indigenous rice cultivar Pokkali. This somaclone yielded 36.3 q/ha under salt-stress conditions as compared to 44 q/ha under normal soil.
- **Aluminium tolerance**: It has long been established that plant species or cultivars greatly differ in their resistance to aluminium stress. In recent years, considerable research has been focused on the understanding of physiological, genetic and molecular processess that lead to aluminium tolerance. Despite the problems encountered in adopting culture media for *in vitro* selection for aluminium resistance, cell lines have been isolated in several species e.g., alfalfa, carrot, sorghum, tomato, tobacco. Aluminium - tolerant somaclonal variants from the cell cultures of carrot were selected by exposing the cells to excess ionic aluminium in the form of aluminium chloride. Jan *et al.* (1997) elicited aluminium toxicity during *in vitro* selection in rice by making several modifications in the media *viz*., low pH, low phosphate and calcium concentrations, and unchelated iron and aluminium along with aluminium sulphate. After selection on aluminium toxic media, callus lines were maintained on aluminium toxic-free media and nine tolerant plants were obtained. Transmission of aluminium-resistance character was identified till the fourth generation.
- **Drought tolerance:** Wang *et al.* (1997) reported a sorghum somaclonal variant line (R111) resistant to drought stress. A novel hybrid was developed by crossing R111

with several sterile lines, indicating that the selection of somaclonal variants is an effective method for creating new source of genetic variation. Drought-tolerant rice lines were obtained by *in vitro* selection of seed-induced callus on a media containing PEG as a selective agent, which simulated the effect of drought in tissue culture conditions.

- **Herbicide resistance:** The growth of weeds in a population of agriculturally-important crop is generally controlled by herbicides. As herbicides have a short residual life, they are repeatedly applied on the crops. Some crops have become susceptible to herbicide due to its repeated application on them. Further, this method of weed control is not economical since herbicides are quite expensive. Efforts have been made to follow *in vitro* approaches to obtain plants expressing tolerance to the herbicides. Exposure of the cell and protoplast cultures to various herbicides induces mutations that may yield tolerant cell lines, which ultimately regenerate into plants. Examples of herbicide-resistant plants regenerated from cell cultures include *Nicotiana tabacum* (against amitrole, bentazone, chlorosulphon, isopropyl N-carbamate, phenmedipham, picloram and paraquat), *Corydalis* (against glyphosate), and *Medicago sativa* (against glufosinate). Some of the resistant traits are inherited as dominant or recessive monogenic alleles.
- **Insect resistance:** Zemetra *et al.* (1993) used *in vitro* selection technique for generation of somaclonal variants resistant to Russian wheat aphid (*Diuraphis noxia*) in wheat. Calli from susceptible wheat cultivar, 'Stephens', were exposed to an extract from aphid. Resistance to aphid was observed in both R2 and R3 generations.
- **Seed quality:** Recently, a variety BIO L 212 of lathyrus (*Lathyrus sativus*) has been identified for cultivation in Central India, which has been developed through somaclonal variation and has low ODAP, a neurotoxin, indicating the potential of somaclonal variations for the development of varieties with improved seed quality.

13.4. GAMETOCLONAL VARIATION

The variation observed among the plants regenerated from cultured gametic cells is termed as 'gametoclonal variation'. Gametoclones can be produced through the culture of either male or female gametic cells, or their derivatives. The most successful and commonly-used method for a wide range of species is culture of anthers, or isolated microspores. Anther and microspore culture technology has been well-established in many plant species for the production of doubled haploid plants and also for *in vitro* screening of microspores against stresses. Since growth of the cells in culture results in single mutations, the cultures of microspores could result in recessive mutations and these can be *in vitro* selected against various stresses. Hence, it is likely that gametoclonal variants can be detected in R0 generation while evaluation of R1 progeny must be completed to detect the full spectrum of variants.

Table 13: Variation among the plants regenerated from gametophytic tissue culture

Crop	Characteristics
Brassica napus	Time to flower, glucosinolate content, leaf shape and colour, flower type, pod size and shape.
Hordeum vulgare	Plant height, yield, fertility, days to maturity
Nicotiana sylvestris	Leaf colour and shape, growth rate, DNA content, repeated sequences of DNA
N. tabacum	TMV resistance, root-knot resistance
	Time of flowering, plant size, leaf shape, alkaloid content, number of leaves
	Altered temperature sensitivity
Oryza sativa	Seed size, seed protein level, plant height, level of tillering waxy mutant
Saintpulia ionantha	RuBSase activity

In crops such as barley, wheat, rice, *Brassica* spp., tobacco, potato etc., gametoclonal variants have been generated, and used to accelerate and enrich the breeding programme. In fact, the first report on selection of disease-resistant mutants by Carlson (1973) utilized haploid tissue of tobacco. Now, there have been several reports of variation in the plants regenerated from gametophytic tissue and some have been documented in Table 13.

■■■■■

14 GERMPLASM CONSERVATION

14.1. INTRODUCTION

The primitive cultivars and wild relatives of crop plants constitute a pool of genetic diversity, which is invaluable for future breeding programmes. However, the existence of these plants has been endangered by the continual introduction of new cultivars. This danger of the erosion of genetic resources was recognized in 1974, when the Consultative Group on International Agricultural Research established the IBPGR with the proposal to develop a global network of genetic resource centres. This was to ensure that the genetic resources of the cultivated plants are collected, conserved, evaluated and made available to the plant breeders anywhere in the world.

Conventionally, germplasms are maintained in the form of seeds because they occupy a relatively-small space and can be stored at low temperature for many years. These seeds are dried to a temperature of -18°C or lower and low humidity. They can be easily dried and packed for transmission to the introduction centres and gene banks. The limitations of this method, however, are:

- Some crops such as cacao, coconut, and mango, large recalcitrant (short-lived) seeds, which lack a dormancy mechanism and cannot bear subjection to desiccation or exposure to low temperature
- The seeds may be destroyed by pathogen and pest attacks.
- Discrete clones cannot be maintained in the form of seeds except for apomictic species.

- It is not applicable to vegetatively-propagated crops such as *Dioscorea, Ipomoea* and potato.

Moreover, the cost of maintaining a large proportion of the available genotypes of the crop plants and endangered species in nurseries or fields is extremely high, and there is a risk of the material being lost as a result of environmental hazards.

The conservation of plants *in vitro* has a number of advantages over *in vivo* conservation. These are as follows:

- *In vitro* culture enables the plant species that are in danger of being extinct to be conserved.
- *In vitro* storage of vegetatively-propagated plants can result in great savings in storage space and time.
- Sterile plants that cannot be reproduced generatively can be maintained *in vitro*.
- It is possible to efficiently reduce the growth *in vitro*, which decreases the number of subcultures necessary.
- If a sterile culture is obtained, often with great difficulty, subculturing *in vitro* is the only safe way to ensure that it remains sterile.

There are two main approaches for *in vitro* storage of germplasm:

(1) Slow growth method and (2) Cryopreservation. Application of these methods invariably depends on the plant material to be used for storage.

1. Slow Growth Method

Cells or tissues can be stored at non-freezing conditions in a slow-growth state rather than at the optimum level of growth. All the cultures in slow growth are static, and therefore, require only shelf space in suitable environments. This approach can be applied to tissue cultures mainly by the following methods:

a. **Temperature:** The tissues can be cultured at low temperatures of 1–4°C. The explants such as callus, shoot-tip cultures are established and maintained for 3–4 wk under standard growth conditions. The explants are then transferred to lower temperatures of 1–4°C for the cultures normally grown at 20–

25°C, and at 4–10°C for the cultures normally grown at 30°C. The ageing of the plant material is slowed down at low temperatures, but it is not completely stopped as in cryopreser vation. Cold storage of *in vitro* derived shoots/plants has been particularly successful in fruit-tree species. Virus-free strawberry plants could be maintained for 6 y at 4°C, provided a few drops of the liquid medium were added to the cultures after every 3 mo. About 800 cultivars of grape plants have been stored for over 15 y at 9°C by yearly transfer to fresh medium. A 16 h photoperiod with reduced lighting (50 lux) may be beneficial in cold storage.

b. **Low-pressure storage and low-oxygen storage:** Attempts have been made to develop LPS and LOS as feasible techniques for future conservation of the cultured plant materials. It is also envisaged to use these procedures as alternatives to cryopreservation and cold storage. In LPS, the atmospheric pressure surrounding the tissue cultures is reduced, resulting in partial decrease of the pressure created by the gases in contact with the plant material. On the other hand, in LOS, the atmospheric pressure (760 mm Hg) is not reduced, but the inert gases (particularly nitrogen) are combined with oxygen to create low-oxygen pressure.

   Low-pressure or hypobaric systems have been found useful for both short-term and long-term storage. Short-term storage is generally used to extend the shelf life of the plant materials such as fruits, cut flowers, vegetables, potted plants and cuttings while long-term storage is used for the materials grown in cultures.

c. **Hormones:** It is possible to use the PGRs such as B-9, CCC and ABA to delay the culture growth. These chemicals cause reduced growth, mainly shorter internodes of the plants.

d. **Osmotic inhibitors:** Cultures can be grown on a medium supplemented with 3–6% (w/v) mannitol or high levels of sucrose as an osmotic inhibitor, which slows down the growth.

2. Cryopreservation

Ever since the publication of 'Life and death at low temperatures' by Luyet and Gehenio in 1940, increased interest has been shown by the biologists to store the living materials at ultra-low temperatures leading to the development of the field of cryobiology. In this method, the plant material is frozen and maintained at a temperature of LN, which is around -196°C. At this temperature, the cells stay in an almost completely inactive state. Cryopreservation is an established method of storing sperm cells in order to use them for artificial insemination during animal breeding programmes. This procedure has been successfully applied for germplasm conservation of a wide range of the plant species such as cassava, pea, chick-pea, rice, wheat, peanut, coconut, oil palm, strawberry, sugarcane and other agricultural crops.

14.1.1. Pre-freezing Treatments

- **Pre-culture:** A brief culture of shoot apices before freezing has proved beneficial in several cases. The first step in cryopreservation is establishment of suitable callus/ cell suspension cultures capable of high-frequency regeneration.
- **Cryoprotectants:** As a result of dehydration of the cells before or during freezing, the intracellular concentration of the solutes increases before freezing of the protoplasm occurs. Cryoprotectants are the substances added to the freezing mixtures to protect the cells from this toxic solution effect. According to Withers and Street (1977), cryoprotectants also prevent the formation of large ice crystals inside the cells. The cryoprotectants are commonly dissolved in the culture medium. For plants, DMSO has proved to be the most effective cryoprotectant. The optimal concentration of DMSO for the cultured cells ranges from 5–8%. However, some of the shoot apices and plantlets may tolerate even higher levels (5–20%). The other compound sometimes used successfully, either alone or in combination with DMSO, is glycerol. It is worth noting that the cell suspensions grown in a nutrient medium supplemented with 10% proline for 3–4 d survived freezing even without further cryoprotection. Plant cells are known to accumulate free proline under water and salt stresses, and it is regarded as a natural protective agent of enzyme and cellular structures under these conditions. The optimal concentration of proline was 10%.

The cultures are placed in sterile plastic vials and in case of liquid cultures; excess of the medium is removed. The cell density, in case of suspension cultures, is suitably adjusted to optimum. Normally, a dilute solution of the cryoprotectant is prepared and added gradually at intervals of 5 min to prevent plasmolysis of the cells. Use of an ice bath while adding a cryoprotectant is beneficial since room temperature may affect the viability of the cells and tissues. After the last addition of the cryoprotectant, there should be an interval of 20–30 min prior to freezing.

14.1.2. Freeze Preservation

Freeze preservation broadly involves four steps: (1) Freezing, (2) Storage, (3) Thawing and (4) Reculture.

1. **Freezing:** Many protocols have been described for the cryopreservation of plant cells. The general practice is to suspend the material in the culture medium and after treating with a suitable cryoprotectant, transfer it to sterile polypropylene ampoules with screw-caps, and freeze by one of the following methods:

   a. **Slow-freezing method:** In this method, the material is frozen at slow-cooling rates of 0.5–4°C per min, starting from 0°C until the temperature reaches -100°C, and finally transferred to LN. The slow-freezing procedure has proved, especially successful with cells from the suspension cultures.

   b. **Rapid-freezing method:** This method is simple as the vials containing the plant material are directly lowered into a tank filled with LN. The temperature decreases rapidly, at the rate of -300–11,000°C/min. Use of dry ice (carbon dioxide) instead of LN exerts a similar effect.

Siebert (1976) and Reed (1992) used the rapid-freezing procedure in which the isolated shoot-apices were frozen by first pouring LN on the freezing vial and then dipping it in a flask of LN. Isolated shoot-apices of *Dianthus caryophyllus* and strawberry, maintained at -196°C for 2 mo, remained viable and regenerated whole plants when cultured. The survival of shoot-tips following this ultra-rapid cooling procedure has been explained on the basis that the critical temperature zone for ice crystal growth is passed so rapidly that the ice crystals of lethal size are not formed inside the cells. The rapid-freezing method has been successfully applied for germplasm conservation of large number of the species.

The water content of the samples is critical in deciding whether to use the slow-freezing or rapid-freezing method. Small specimens with low water content are appropriate materials for rapid-freezing method. Shoot-apices of strawberry and *Solanum goniocalyx* are the materials with small cells potentially able to regenerate plants after rapid freezing. The main drawback in following this procedure is that some plant materials do not survive rapid freezing by direct immersion in LN, and therefore, stepwise freezing becomes necessary.

c. **Stepwise-freezing method:** This method combines the advantages of both slow-freezing and rapid-freezing procedures. The plant material is cooled stepwise (ca 1–5°C/min) to an intermediate temperature, maintained at that temperature for 30 min, and then rapidly cooled by plunging it into LN. The stepwise-freezing method can be used for freeze-preservation of a wide range of plant materials, which include shoot-apices, buds and suspension cultures. In the initial slow-freezing, ice is formed outside the cells and the unfrozen protoplasm of the cells loses water due to the vapour deficit pressure between the supercooled protoplasm and the external ice. Compared to rapid-freezing method, strawberry shoot-tips showed enhanced survival rate (60–80%) using this method (Sakai, 1997).

d. **Dry-freezing method:** Materials dehydrated by drying in an oven or under vacuum demonstrate remarkable resistance to cryogenic damage. The basic idea for the dry-freezing method originated from the fact that non-germinated dry seeds are able to survive freezing at superlow temperature in contrast to water-imbibing seeds, which show susceptibility to cryogenic injuries. Similarly, dehydration of the cells under vacuum also leads to a better survival of plant organs after freezing at -196°C. However, there is a dehydration optimum, which varies according to species and tissues.

e. **Vitrification:** Water content in the cells of the material to be preserved is critical in deciding whether to use slow-freezing or rapid-freezing methods. It is essential to avoid intercellular freezing to maintain viability of the cells with more water content after cooling to the temperature of LN.

This can be accomplished either by drying or vitrification. The drying process is not always successful as there is a dehydration optimum, which varies according to species. Another potential procedure that prevents cells from cryogenic injuries and successfully applied to most of the plant materials for *in vitro* conservation is vitrification.

Vitrification is an effective mechanism against freezing wherein a highly-concentrated cryoprotective solution supercools to very low temperatures and solidifies into metastable glass without undergoing crystallization. The metastable glass is exceedingly viscous and stops molecular diffusion inside the cells. In the vitrification method, cells or meristems are sufficiently dehydrated with a highly-concentrated vitrification solution at 25°C or 0°C without causing injury prior to immersion in LN.

f. **Encapsulation/Dehydration:** Cryopreservation by vitrification requires careful control of the procedures for dehydration and cryoprotectant permeation in order to prevent injury by chemical toxicity or excess osmotic stress during dehydration. In the encapsulation/dehydration technique, the cells/tissues are trapped into calcium alginate beads followed by incubation in 0.85 M sucrose (as sole source of cryoprotectant) for 4–16 h and rapid freezing in LN. The beads are air-dried for 3–4 h in a laminar flow chamber prior to immersion in LN.

2. **Thawing:** Rapid thawing of the material frozen at -196°C is achieved by plunging it into water at 37–40°C, which gives a thawing rate of 500–750°C/min. After about 90 s, the material is transferred to an ice bath and maintained there until recultured or its viability is tested. Slow thawing at room temperature has generally proved fatal. Rapid thawing seems to protect the cells from the damaging effects of ice crystal formation, which may occur during slow warming. The principle involved is the same as in the rapid-cooling method of freezing plant material. However, if the water content of the cells has been reduced to an optimum level prior to freezing, the thawing rate is less critical.

3. **Reculture:** Thawed materials are washed several times to remove cryoprotectants so as to avoid any deplasmolytic injury to the cells. The washed material is then recultured in a fresh

culture medium following the standard procedures. The frozen and thawed cells of *Zea mays* showed a faster recovery and higher survival rate if they were recultured in the presence of the surrounding medium.

The plant material frozen at -196°C may exhibit some special requirement for better survival when recultured. Shoot-tips from frozen seedlings of tomato directly develop into plantlets only if the medium was supplemented with GA_3. In the absence of GA_3, the apices callused, followed by the differentiation of adventitious shoots. The GA_3 was not required by the control (non-frozen) shoot-apices to directly develop into plantlets. The GA_3 (10 mg/l) in the recovery culture medium also improved the survival rate of cryopreserved zygotic embryos of coffee. The overall survival of the frozen and thawed plantlets of carrot was greatly enhanced by AC.

14.2. DETERMINATION OF CELL SURVIVAL

- **Fluorescein diacetate staining:** This is based on the ability of living cells to get stained with FDA and showing fluorescence under UV light. Dead cells do not show any fluorescence. About 0.1% FDA is mixed with the cell culture and allowed to stand for 5–10 min. They are first observed under normal light in a microscope; cell number is counted and then they were observed under UV light. The percentage of the living cells was estimated by noting the surviving cells by their fluorescence.
- **Triphenyltetrazolium chloride method:** Cell survival is estimated by the amount of formamide (pink colour) produced by the reduction of TTC by the living cells.
- **Growth measurements:** This includes mitotic index, increase in cell number or culture volume, dry and fresh weight increase and PE. These are also used to indicate the living and active state of the cultures.

14.3. ACTION OF CRYOPROTECTANTS

The actual mechanism is not clearly understood. Various views include:

- They act as antifreeze agents by reducing the residual concentration of intracellular salts (Lovelock, 1953), or by diluting the toxic solutes or by preventing ice crystal formation.

- Levitt's (1962) Sulphydryl–disulphide hypothesis: Dehydration induces intermolecular SS bond formation and the cryoprotectants prevent this.
- Rowe (1966) suggested that they interact directly or indirectly with the cell membrane to stabilize the water-lipid-protein complex tertiary structure. Thus, the biological and physiological integrity of the membrane is protected.

14.4. FACTORS AFFECTING REVIVAL OF FROZEN CELLS

A number of factors were found to influence the successful cryopreservation. They can be broadly grouped into five major classes:

1. Nature of culture

a. Material taken from fresh or just subcultured calli or young suspension cultures in exponential growth phase were found to tolerate the freezing process. Large thick-walled cells with scanty cytoplasm or highly-vacuolated cells could not survive the freezing injury due to accumulation and bursting of the cells within the vacuoles.

b. Cells or tissues with high water content were forced to be more susceptible. Therefore, spores and seeds can be easily subjected to cryopreservation.

c. Small colonies or clumps of the cells with dense cytoplasm survive better than the individual or single cells.

d. Greater the cell density or biomass per unit volume of the medium, greater the tolerating capacity to the freezing temperatures.

e. If the mother plant from which the cultures were established has originally frost/winter hardiness, the derived cultures might be more tolerant to freezing.

2. Cryoprotective agents

Properties of a good cryoprotective agent are:

a. Low molecular weight

b. Easily-miscible nature in the solvent

c. Non-toxic at lower concentrations

d. Washability with water and possibility of easy removal from the cells

e. High penetrability into the cells

The DMSO was found to be ideal among the several chemicals such as glycerol, ethylene glycol, diethylene glycol, propylene glycol, dimethyl acetamide, PVP, and hence, is frequently used. In some cases as in *Ipomoea*, a combination of cryoprotectants was used. The type and concentration of the cryoprotectants vary with species.

3. Cooling

Rate of cooling is one of the critical parameters. Direct and sudden lowering of the temperature may disturb physiological metabolic pathways, and might cause extracellular and intracellular formation of ice crystals, thus causing dehydration. The formation of intracellular ice crystals is mainly responsible for freeze injury. To prevent this, cooling must be slow and gradual to allow all freezable water to flow out of the cells.

4. Storage temperature

The temperature should be low for long-term storage of the cells to stop all metabolic activity and prevent biochemical injury. Different temperatures are used for different species and it also depends upon the time of storage.

5. Thawing

Rapid warming to 37ºC was found to be better because it prevents the recrystallization or growing of crystal size, which generally happens during slow warming.

14.5. APPLICATIONS

- **Conservation of genetic material:** Large numbers of plant species have been successfully maintained by means of cryopreservation of cultured embryos, tissues, cells or protoplasts. A list of cryopreserved tropical plants is given in Table 14.

Table 14: A list of tropical plants cryopreserved in various forms

Cell suspensions	**Meristems**
Berberis dictyophila	*Oryza sativa*
Capsicum annuum	*Zea mays*
Glycine max	**Protoplasts**
Nicotiana plumbaginifolia	*Glycine max*
Nicotiana tabacum	*Nicotiana tabacum*
Callus	*Capsicum annuum*
Gossypium arboreum	*Coffea arabica*
	Daucus carota
Pollen embryos	*Arachis hypogaea*
Arachis hypogaea	*Citrus* spp.
Cicer arietinum	*Nicotiana tabacum*
Solanum tuberosum	*Atropa belladonna*
Somatic embryos	
Citrus sinensis	
Saccharum sp.	
Capsicum annuum	
Oryza sativa	

- **Freeze storage of cell cultures:** A cell line to be maintained has to be subcultured, and transferred periodically and repeatedly over an extended period of time. It also requires much space and manpower. Freeze preservation is an ideal approach to suppress cell division and to avoid the need for periodical subculturing.
- **Maintenance of disease-free stocks:** Pathogen-free lines can be cryopreserved and propagated when required. Cryopreserved material can also be used for exchange of material between countries since it is pathogen-free.
- **Cold acclimatization and frost resistance:** Tissue cultures would provide a suitable material for selection of cold-resistant mutant cell lines, which could later differentiate into frost-resistant plants.

- **Use in gene banks for preservation of rare genomes:** Rare and endangered plants can be preserved for long periods by using their seeds for cryopreservation. Several chromosomal and genetic abnormalities can occur in long-term tissue cultures and these may not be of natural occurrence in intact plants. These may contribute to somaclonal variation. Such uncommon cell/tissue lines can be preserved for posterity.
- **Conservation of genetic uniformity:** Tissue or cell cultures with no genetic changes can be stored and made available to the breeders and others as standard stock to provide uniform experimental material.

■■■■■

15 PRODUCTION OF SECONDARY METABOLITES THROUGH TISSUE CULTURE

15.1. INTRODUCTION

Plants have been the richest source of several products, which are of immense use to the human beings. Many higher plants are major source of natural products used as pharmaceuticals, agrochemicals, flavour and fragrance ingredients, food additives and pesticides. Noteworthy among them and next to those of food value are the substances of medicinal importance. The search for new plant-derived chemicals should be a priority in current and future effort towards sustainable conservation. The metabolic pathways that are essential for sustaining life and organization of plant body such as synthesis of starches, proteins, nucleic acids and fats are regarded as primary metabolism. The other series of reactions that enable the plant to interact successfully with the biotic and abiotic environments are considered to constitute secondary metabolism. Often these two – primary and secondary metabolic pathways – are closely interconnected in a plant system. About 20,000 compounds (including alkaloids, steroids, peptides, etc) are known to occur as the products of secondary metabolism and a majority of these have medicinal or pharmaceutical importance.

The difficulties that are increasingly encountered in procuring ample supply of raw plant material because of drastic decrease in natural resources due to human disturbance to the environment, ruthless exploitation of the available resources, increasing labour costs, and technical as well as economic difficulties in cultivating wild plants, have prompted the adaptation of *in vitro* technology for the commercial production of the substances of medicinal importance. Besides providing an alternative technology to bypass the above difficulties, the plant tissue culture (used in a broad sense to include cell, tissue and organ culture) offers some other advantages (Tabata, 1977):

- Useful compounds could be produced under controlled environmental conditions independent of climatic changes or soil conditions.
- Cultured cells would be free of microbes and insects.
- The cells of any plants, tropical or alpine, could easily be multiplied to yield their specific metabolites.
- Automated control of the cell growth and rational regulation of the metabolic processes leading to improved productivity of the desired products

In addition to these, the *in vitro* technology also facilitates:

- Novel means of conserving the genetic diversity of the germplasm of medicinal plants through cryopreservation
- Production of novel compounds through biotransformation, somatic hybridization and selective gene transfer through recombinant DNA technology

15.2. TYPES OF CULTURES

The beginning of the era of using plant cell, tissue and organ cultures for the production of secondary metabolites on a commercial scale can be traced to the work of West and Mika (1957) dealing with the synthesis of atropine from root-derived callus cultures of *Atropa belladonna*. Since then, a variety of cultures have been employed for the production of active principles of medicinal value. Choice of the system mostly depends on the site of biosynthesis/accumulation of the compound as well as the nature of the compound. The following are the different types of cultures used for the extraction of various medicinal substances.

Types of plant cell culture systems used for secondary metabolite extraction can be broadly categorized into organized and unorganized cultures.

1. Unorganized cultures

a. **Cell suspension cultures:** When callus pieces are agitated in a liquid medium, they tend to break up and these are then called cell suspension cultures. Friable callus goes easily into the suspension. Cell aggregates are removed by sieving. Plating of single cells and small aggregates are done; only viable cells will grow and can be reintroduced into the suspension. 2,4-D, low cytokinin, semi-solid medium, enzymatic digestion with pectinase and blending are some of the requirements for the suspension cultures. Suspensions are much easier to bulk up than callus since there is no manual transfer or solid support. Cell suspensions have been the most preferred type of cultures for the production of valuable medicinal compounds. In some cases, the spectrum of compounds produced from cell cultures is generally very low. A very close correlation between the growth of the cultures and yield of the product has been observed. The product accumulates through the growth cycle, and thus, the product and biomass show a closer relationship.

The behaviour of the cell lines of a culture differs with the timing of secondary metabolite production during culture cycle. Cultures of *Catharanthus roseus* were observed to accumulate alkaloids during the active growth of culture, while in some other cases, the synthesis occurred during the stationary phase. In cultures where the accumulation of the product occurs simultaneous with the growth, it is necessary to maintain the cells in the proliferation stage to obtain higher yields of the product.

Table 15: Industrially-developed plant cell suspension cultures

Product	Source	Company	Country
Shikonin	*Lithospermum erythrorhizon*	Mitsui Petrochemical	Japan
Berberine	*Coptis japonica*	Nitto Denko	Japan
Biomass	*Panax ginseng*	Toyobo	Japan
Peroxidase	*Raphanus* species	Kanebo	Japan
Geraniol	*Geranium* species	Natterman	Japan
Rosmarinic acid	*Coleus blumei*		
Digoxin	*Digitalis lanata*	Boehringer	Japan

Source: Mor, 1991

Use of cell suspension cultures for the production of secondary metabolites was observed to have certain limitations. The cell lines failed to produce certain valuable alkaloids on high yields in the poorly-differentiated cultures as in the case of the production of vinblastine and vincristine, which are antineoplastic agents produced from *Catharan-*

thus roseus, and morphine and codeine from *Papaver somniferum*. Similarly, diosgenin from *Dioscorea deltoidea*, sanguinearine – an antibiotic from *Papaver somniferum*, paclitaxel (Taxol) – an anticancer drug from *Taxus brevifolia* are highly useful compounds, but their yields are very low.

The difficulties in producing secondary products using suspension cultures, such as rapid loss or very low quantities of the product produced, led to the alternative use of cell immobilization techniques. Immobilization of the cells has received due attention based on the advantages such as:

- Easy recovery of the cells and product from the medium, which enables a continuity of the process
- Increased stability of the cells and accumulation of the secondary products

In many cases, secondary product synthesis, typical of intact plant, cannot occur in rapidly-growing undifferentiaed cell cultures, but requires some degree of morphological or biochemical differentiation and slow growth, and needs two-stage cultural system.

Table 16: Product under development from plant tissue culture (Heble, 1996)

Product	Plant species	Biological activity
Taxol	*Taxus brevifolia, T. baccata*	Ovarian cancer
Camptothecin	*Camptotheca acuminata*	Anticancer
	Nothapodytes foetida	
Forskolin	*Coleus forskohlii*	Cardiotonic
Artemesinin	*Artemesis annua*	Malaria
Castanospermine	*Castanospermum australe*	AIDS
Hypericin	*Hypericum esculentum*	AIDS

b. **Callus cultures:** In the very first attempt to produce a secondary metabolite under *in vitro* conditions, callus cultures of *Atropa belladonna* have been used. However, several subsequent workers involved cell suspension culture due to its faster growth rate. However, callus cultures are still in vogue as in *Artemisia annua, Camptotheca acuminata, Catharanthus roseus, Dioscorea pentaphylla* etc. probably due to the ease in their establishment and handling. List of the secondary metabolites produced through callus cultures is given in Table 17.

Requirement for support ensures that scale up is limited.

Table 17: Secondary metabolites produced through callus cultures

Compound	Species	Reference
Allin	*Allium sativum*	Malpatak and David (1986)
Saikosaponin	*Bupleurum falcatum*	Wang and Huang (1982)
Canavanine	*Canavalia ensiformis*	Ramirez *et al.* (1992)
Pyrethrins	*Chrysanthemum cinerariaefolium*	Rajasekhara *et al.* (1992)
Caffeine	*Coffea arabica*	Wall *et al.* (1991)
Planotol	*Croton sublyratus*	Monmoto and Murai (1987)
Cryptosin	*Cryptolepis buchanani*	Venkateswara *et al.* (1987)
Sterols	*Eucalyptus tereticornis*	Venkateswara *et al.* (1986)
Triterpenes	*Glycyrrhiza glabra*	Ayabe *et al.* (1990)
Glycosides	*Gymnema sylvestre*	Gopi and Vatsala (2006)
Alkaloids	*Hyoscymus niger*	Yamada and Hashimoto (1982)
Saponins	*Panax ginseng*	Asaka *et al.* (1993)
Morphine	*Papaver somniferum*	Furuya *et al.* (1972)
Reserpine	*Rauvolfia serpentina*	Gerasimenko *et al.* (2001)

2. Organized cultures

The lacunae met with the cell suspension and callus cultures such as non-expression of relevant pathways in undifferentiated cells as they are under developmental control, lack of mechanisms for product removal, lack of storage sites etc. have been overcome to some extent by using organized cultures such as root and shoot cultures, that resulted in higher yields of secondary products.

a. **Hairy root cultures:** Two main types of root cultures *viz.*, the untransformed excised root cultures and transformed hairy root cultures have been recognized. Eventhough the former type provided good source system for the accumulation of secondary compounds, the exploitation of the latter type seems to have greater potential for secondary metabolite production.

For establishing hairy root cultures, the plants are infected by *Agrobacterium rhizogenes,* which includes hairy root disease by the transfer of T-DNA from Ri plasmid into the plant genome. Genes for auxin synthesis and sensitivity are engineered into the plant cells leading to gravity-insensitive mass root production. These transformed roots are genetically stable and retain their biosynthetic activity for longer periods. The growth in these roots is comparable to that of the cell suspensions and is faster than that of the normal roots. Genes required for the expression of biosynthetic pathways can be transferred and expressed in hairy root cultures.

A wide variety of compounds have been synthesized through the transformed or hairy root cultures. The secondary metabolites produced from hairy root cultures exhibit a number of advantages over the cell cultures such as:

- The relative ease of culturing
- Simple media they require
- The genetic stability
- Rapid growth with greater accumulation of biomass in short period
- Steady level of product synthesis over a relatively long period and the production of new compounds, which are absent in cell cultures, but present in roots of intact plants

Aggregation and shear sensitivity are the major problems for scale up.

b. **Shoot cultures:** Some of the substances of medicinal value are specifically synthesized in commercially-viable quantities by only the shoots. In such cases, they have to be extracted only from the shoot cultures as in the case of artemisin from *Artemisia*, galantamine from *Narcissus* etc. A culture producing mass of shoots may be produced by adventitious shoot formation under the conditions of high cytokinin. The shoot cultures may be much higher yielding for high-associated products. Illumination is a problem for scale up. When the metabolite is not produced in callus and suspension cultures, but only in intact plant, then the shoot cultures have to be carried out. Organization for enhancement of the production of higher secondary metabolites can be made through medium manipulations.

15.3. BIOREACTORS

Large-scale cultivation of plant cells is done in bioreactors and the stirred-tank type is the commonly-used design (Scragg, 1998). Large-scale (50,000 l) commercial fermentations were possible for shikonin and berberine production. This type of reactor uses mechanical agitation by means of a rushtone turbine. It stirs and improves the air supply by producing turbulent high-shear conditions. The high shear produced by this kind of impeller can cause cell breakage in some suspension cultures. Culture of such shear-sensitive cells is done in air-lift type of bioreactors. These were first used for the production of single cell proteins and used a stream of air sparged into the base of the bioreactor, which results in both aeration and mixing of the culture. The air-lift type of reactor contains two interconnected unequal compartments. Here, sparging of the air into one sector results in reduced density and consequent upflow of liquid that, in turn, leads to an inflow of liquid from the other section, thus establishing a flow.

The types of reactors for plant cell suspension cultures include modified stirred-tank, air-lift, air-loop, bubble-column and rotating-drum reactor. Details of these reactors are given in Chapter 5 – Cell and Suspension Culture.

15.4. CELL IMMOBILIZATION

Large-scale culture of immobilized cells uses fluidized-bed reactor and packed-bed reactor. Usage of these different types is dependent upon the method and form of immobilization used (such as gel entrapment, polyurethane entrapment or membrane entrapment). Similarly, use of other types of bioreactors such as bubble-column type, spin-filter type etc. are used for the organized cultures.

In immobilized cell systems, the cells are physically entrapped in a gel and the products can diffuse from immobilized cells by entrapment in calcium alginate, potassium carrageenan or in agarose beads. Advantage of this method is automatic physical separation of the product from cells, reuse of extensive biomass that results in continuous operation and stabilization of biocatalyst (producing cells). Immobilized cell systems include immobilized hairy root cultures. In suspension cultures, the substances used for immobilization include synthetic polymers (polyacrylamide, epoxyresin, polyurethane), carbohydrates (cellulose, agar, agarose, chitosan, alginate) and proteins (collagen, gelatin, fibrin). The list is rapidly growing and it appears possible to

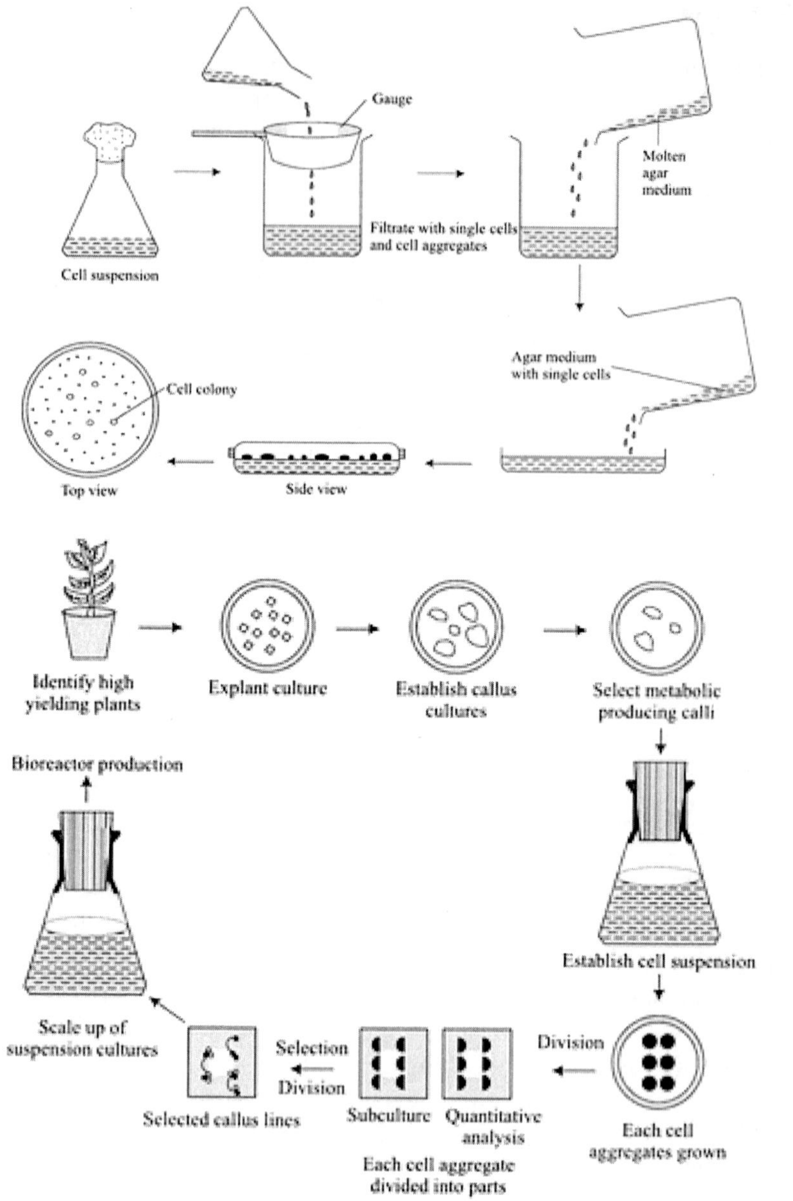

Fig. 46 Production of secondary metabolites

immobilize any plant cell line with preserved viability. A differeent approach is to allow the cells to become entrapped within the spaces of a preformed matrix such as nylon, polyurethane and metal mesh. Viability tests should be carried out after immobilization. In immobilized cell systems, continuous cultivation can only be applied to extracellular products. Products normally stored within the cells leak out into the medium. If not, the induced release is performed. Secondary products are often stored within the vacuoles, and then permeabilizing agents such as DMSO are needed for release of the compound.

15.5. IMPROVING THE PRODUCT QUALITY AND QUANTITY

15.5.1. Elicitors

The recognition that certain specific secondary metabolite products such as phytoalexins are produced by the plants, which are active against microorganisms, has led to the concept of using such stimulators for *in vitro* cultures also. Such compounds have been described as 'elicitors' by Keen *et al.* (1972). There are two kinds of elicitors:

1. Abiotic elicitors
2. Biotic elicitors

Abiotic elicitors are physical or chemical in nature e.g., UV radiations, alkalinity, osmotic pressure or heavy metal ions etc. Biotic elicitors are complex culture homogenates of fungal or bacterial origin. The substances used as elicitors or the microorganisms producing elicitors may be pathogenic or non-pathogenic. Fungal elicitors have been shown to modify the synthesis of intermediates leading to various cell wall components (Funk *et al.,* 1990). An elicitor from *Botrytis* was noticed to trigger the synthesis of a potentially-important alkaloid, sanguinarine, in large quantities. Yeast elicitor, a specific carbohydrate fraction, was shown to induce significant increase in the yields of berberine and benzophenanthridine alkaloids from *Thalictrum rugosum* cultures, and of the latter category of substances in *Eschcholzia californica* also. The abiotic elicitors include physical and chemical stresses such as UV radiation, exposure to cold or heat, ethylene, fungicides, antibiotics, salts of heavy metals, salinity etc. (Yeoman and Yeoman, 1996).

Table 18: List of potential abiotic elicitors

Metabolites	System	Elicitors
Ajmalacine	*Catharanthus roseus*	Diethylaminoethyl dichloro phenyl ether Catharanthine
Echinofuron	*Lithospermum erythrorhizon*	Activated carbon
Glyceollin	*Glycine max*	Laminarin, digitonin
Flavonoid glycosides	*Haplopappus gracilis*	UV light
Berberine	*Thalictrum rugosum*	Ethephon
Caffeine	*Coffea arabica*	Ethephon
Pisatin	*Phaseolus vulgaris*	Glutathione
Kievitone	*Phaseolus vulgaris*	Denatured RNAse

Table 19: List of potential biotic elicitors

Metabolite	System	Elicitor (mycelial/cell extracts or filtrates)
Thiophene	*Tagetus patula*	*Fusarium conglutinane*
Sanguinarine	*Eschscholzia californica*	*Fusarium oxysporum*, *Penicillium* sp.
Diosgenin	*Dioscorea deltoidea*	*Rhizopus arrhizus*
Benzofurons	*Ageratina adenophora*	*Saccharomyces cerevisae*
Tropane alkaloids	*Datura stramonium*	*Phytophthora megaspora*
Ajmalicine	*Catharanthus roseus*	*Micromucor isobellina*
Furanocoumarine	*Petroselenium hortense*	*Alternaria carthanis*
Medicarpin	*Cicer arietinum*	*Ascochyta rabies*
Phytuberine	*Nicotiana tabacum*	*Pseudomonas solanacearum*
Capsaicin	*Capsicum annuum*	*Gliocladium deliquescens*
Anthraquinone	*Morinda citrifolia*	*Aspergillus niger*
Anthocyanins	*Daucus carota*	*Aspergillus, Rhizopus*

15.5.2. Biotransformation

Addition of the compounds (not necessarily present in the species) to the media as precursors or to be transformed to economically-important alkaloids or other compounds is known as biotransformation.

Another interesting aspect regarding the use of *in vitro* cultures for the extraction of secondary metabolites is the possibility of feeding the precursor or one of the intermediates in a metabolic pathway and its subsequent conversion to the final product, ultimately resulting in its increased yield. This phenomenon of precursor feeding leading to its subsequent conversion into other compounds by the living cells is commonly referred to as biotransformation. Bennett and Heftmann (1965) demonstrated the uptake and conversion of labelled cholesterol to diosgenin and kryptogenin by the seedlings of *Dioscorea spiculiflora*. However, digitoxin was the first cardiac glycoside used for the biotransformation experiments by plant cell cultures (Stohs and Staba, 1964).

Usually, the exogenous feeding of the precursor for the purpose of biotransformation is done beyond the stages of metabolic pathway, where the common points for primary and secondary metabolisms do not occur. It was observed that applying the precursor at some distance from the product may result in lesser quantum of increase than when the precursor is applied at a step closer to the product (Lindsey and Yeoman, 1983). Some of the major problems associated with the biotransformation technique are cost of the precursor relative to that of the product, inability of the cultures for the precursor uptake, toxic nature of the precursor at the concentration levels used for feeding, exact timing of precursor feeding and the possibility of exogenous precursor being diverted to the pathways other than the desired one, thus reducing the efficiency of desired bioconversion. Transformation process may occur by oxdiation, reduction, hydroxylation, glycosylation, esterification, methylation, demethylation, isomerization and epoxidation.

Biotransformation of progesterone was achieved by suspension cultures of *Digitalis purpurea* cultured cells (Furuya *et al.,* 1973). The cells of *Digitalis lanata* are able to transform digitoxin to digoxin affected by enzymes present in the cell. Digitoxigenin was converted to non-glycosidic compounds both by *Digitalis* and mint tissue cultures.

Digitalis purpurea cultures specifically convert digitoxin to purpurea glycoside A, B, gitoxin and reduced progesterone (Furuya *et al.,* 1973). Tissue cultures of *Digitalis mertonensis* will utilize cholesterol to produce carbenolides.

Biotransformations are performed in the plant cell cultures of *Choisya ternata* for 5-formyl ellipticine production, *Galium mollugo* for

2-succinyl benzoate anthraquinones hydroxylated derivatives production, *Solanum tuberosum* for solavetivone production, *Nicotiana tabacum* for carvaxone production from carvaxiome and *Citrus* species for nootkatone production from valencene.

15.5.3. Somatic Hybrids

A few somatic hybrids were also examined for their secondary metabolite production. Theoretically, the attraction for somatic hybrids lies in the possibility that the total spectrum of active principles in the two parents could be expressed in the hybrids due to genetic complementation. For example, in the hybrid clones derived from the fusion of protoplasts of *Hyoscyamus muticus* and *H. albus*, complementation was noticed for floral characters, peroxidase zymogram profile, and for increased quantity for tropane alkaloids over both the parental quantities.

15.5.4. Somaclonal Variation

Spontaneous development of genetic variability obtained from tissue culture is called somaclonal variation (Larkin and Scowcroft, 1981). Majority of the studies reported on somaclonal variation noticed an increase (both quantitative and qualitative) in the spectrum of variability for the active principles as in the cases of total alkaloid content in the regenerants of *Atropa belladonna*, cardenolides in the protoclones of *Digitalis lanata*, hypericin in the somaclones of *Hypericum perforatum*, stephanin in cell clones of *Stephania*, and in total alkaloid spectrum in the protoclones of *Catharanthus roseus* etc. In addition to these observations on the increasing variability, higher yields of the specific active principles as compared to the *in vivo* quantities were recorded for diosgenin in embryo clones of *Costus speciosus*.

15.5.5. Use of Molecular Markers

These molecular markers (RAPD and RFLP) have been used to tag the sequence associated with the production of a specific secondary metabolite to enable the identification of species producing similar compounds. In addition to this, these markers have also been used to analyse the genetic diversity in the production of a specific compound. RAPDs have been mostly used as the markers, though in few cases such as *Duboisia* and *Plantago*, RFLPs have been made use of.

15.5.6. Gene Cloning and Transgenosis

Using the technique of gene cloning, genes have been isolated, characterized and cloned into desired host system. Genes related to alkaloid biosynthetic pathways have been cloned from a number of medicinally-important plant systems. In *Catharanthus roseus*, the promoters for NADPH-cytochrome reductase gene have been characterized (Lopes *et al.,* 1997). This offers the possibility to express the cloned genes into the desired culture types. Consequent to the successful cloning of genes/related sequences responsible for the production of the active principles, attempts have also been made in few cases to introduce these cloned genes and to develop transgenic plants with increased abilities either for the production of the original compound or for the synthesis of a novel compound. In majority of the cases reported, the genetic transformation was mediated by Ti plasmid (*Atropa belladonna*, *Solanum lacinatum*, *Tabernemontana*, and *Nicotiana*), whereas in *Catharanthus*, the technique of particle bombardment was used.

Table 20: Secondary metabolites of medicinal importance isolated from *in vitro* cultures

Plant species	Culture type	Substance isolated
Artemisia annua	Callus	Artemisinin
	Cell Suspension	Artemisinin, artemisin, sesquiterpene
	Hairy root culture	Artemisinin
Cassia tora	Cell suspension	Germichrysone
Catharanthus roseus	Callus	Ajmalicine, serpentine
	Cell Suspension	Several alkaloids
	Hairy root culture	Several alkaloids
	Cell suspension with homogenate of *Botrytis* species as elicitor	Catharanthene
Datura innoxia	Cell suspension	Tropane, scopalamine
	Hairy root culture	Hyoscyamine, scopalamine
Digitalis lanata	Cell suspension	Cardenolide
	Shoot culture	Cardenolides, digitoxine

Plant species	Culture type	Substance isolated
Lithospermum erythrorhizon	Cell suspension	Shikonin
	Hairy root culture	Shikonin
Papaver somniferum	Cell suspension	Sanguinarine, morphinane,
	Shoot culture	Thebiane, morphine, codeine
	Root culture	Morphinane, morphine, codeine
	Hairy root culture	Sanguinarine
	Cell suspension with homogenate of *Botrytis* or *Pythium aphani-dermatum* species as elicitor	Sanguinarine
		Sanguinarine
Rauvolfia serpentina	Cell suspension	Reserpine
	Hairy root culture	Several indole alkaloids
Withania somnifera	Shoot culture	Withanolide

Table 21: Important secondary compounds produced through tissue cultures

Compound	Species	Uses
Stevioside	*Stevia rebaudiana*	Sweetener
Jasmine	*Jasminum* sp.	Perfume
Codeine	*Papaver* sp.	Analgesic
Atropine	*Atropa belladonna*	Muscle relaxant
Morphine	*Papaver somniferum*	Analgesic
Quinine	*Cinchona officinalis*	Antimalarial
Quinones	*Cinchona succirubra*	Antimalarial
Nicotine	*Nicotiana tabacum*	Insecticidal
	Nicotiana rustica	
Catharanthine	*Catharanthus roseus*	Anticancerous
Shikonin	*Lithospermum erythrorhizon*	Cosmoceutical
Saikosaponin	*Bupleurum falcatum*	Photoprotective
Diosgenin	*Dioscorea deltoidea*	Antifertility

Table 22: Production of pharmaceutically and industrially-important compounds from cultured cells

	Compound	**Species**
1.	Nicotine	*Nicotiana tabacum*
2	Anthocyanin	*Euphorbia milli*
3	Anthraquinone	*Morinda citrifolia*
4	Shikonin	*Lithospermum erythrorhizon*
5	Phenolic compounds	*Acer pseudoplatanus*
6	L Dopa (α L – hydroxyphenylalanine)	*Stizolobium hassjoo*
7	Ferruginol	*Salvia miltorhiza*
8	Sanguinarine	*Papavar somniferum*
9	Ajmalacine and Serpentine	*Catharathus roseus*
10	Rosmarinic acid	*Coleus blumei*
11	Diosgenin	*Dioscorea detoidea*
12	Berberine	*Coptis japonica*
13	Ginseng saponins	*Panax ginseng*
14	Anthocyanin	*Daucus carota*
15	Rosmarinic acid	*Anchusa officinalis*
16	Biotin	*Lavandula vera*
17	Ubiquinone 10	*Nicotiana tabacum*
18	Capsaicin	*Capsicum frutescens*
19	Solasodine	*Solanum xanthocarpum*
20	L Dopa	*Mucuna pruriens*
21	Diacetyllantoside C	*Digitalis lanata*
22	Anthraquinone	*Cassia tora*
23	Thebaine	*Papaver bracteatum*
24	Benzophenanthridine alkaloids	*Eschcholzia californica*

15.6. COMMERCIAL PRODUCTION

Mitsui Petrochemical Industries (Tokyo) is commercially producing shikonin, used both as a dye and pharmaceuticals from *Lithospermum erythrorhizon* cell suspension cultures. Shikonin is also used as herbal lipstick.

Similarily, it has been possible to establish cell cultures for industrial production of berberine from *Coptis japonica* and ginseng extracted from roots of *Panax ginseng*.

Plant cell cultures are known to produce secondary metabolites hitherto not known to occur in the intact plant, as for instance, *Andrographis paniculata* (sesquiterpenes), *Morinda jasminoides* (anthraquinones), *Ruta graveolens* (furocoumarin).

15.7. CHARACTERISTICS OF PLANT CELLS IN SUSPENSION CULTURE

- Cells in suspension culture are large (10–100 μM long).
- They tend to occur in aggregates.
- They are shear-sensitive.
- They are slow growing.
- They are easily contaminated.
- They have low oxygen demand (kla of 5–20).
- They will not tolerate anaerobic conditions.
- They can grow to high cell densities (>300g/l fresh weight).
- They can form very viscous solutions.
- Oxygen demand is proportional to cell density and shear rate is proportional to viscosity.

15.8. FACTORS AFFECTING SECONDARY METABOLITE PRODUCTION IN CELL SUSPENSION CULTURES

- Secondary metabolite production in suspension cultures is affected by the nutrient composition of the medium, which include phosphate, nitrogen and sulphate.
- It is also affected by carbohydrates and growth regulators such as 2,4-D.
- An exogenous supplement of the biosynthetic precursors may increase the yield.

- Many environmental factors affect the production of secondary metabolites. Production of flavonoids, polyphenols, carotenoids and plastoquinones are stimulated by light. White/blue light may inhibit alkaloid biosynthesis. Temperature is a limiting factor in lipid/alkaloid biosyntheis in tobacco and *Papaver* species.
- Heterogeneity of the cultured plant cells is another limiting factor.
- Production of side-products should be prevented.

15.9. WAYS TO INCREASE PRODUCT FORMATION

Appropriate cell lines should be selected to increase the product formation. These cells are generally selected from producing part of the intact plant. The medium is modified for growth and product formation. Precursors or intermediates are fed into the medium for bioconversion. Plant-like conditions such as immobilization are created.

Selection strategies include positive, negative, visual and analytical screening. In positive selection, a toxic compound such as hydroxyproline or kanamycin is added to the medium. Only those cells that are able to grow in the presence of selective agent give colonies. Plate out and pick off the growing colonies. It is possible to select one colony from millions of the plated cells in a days work. A strong selection pressure is needed and some escapes may be there. In negative selection, an agent such as chlorate/BudR, which kills the dividing cells, is added to the medium. Plate out the cells, leave for a suitable time, wash out the agent and then put them on the growth medium. All the cells growing on the selective agent will die, leaving only the non-growing cells now grow. This method is useful for selecting auxotrophs. Visual selection is useful only for coloured or fluorescent compounds such as shikonin/ berberine/ some alkaloids. Plate out about 50,000 cells per plate and after culture pick off coloured cells. It is possible to screen about 1,000,000 cells in a day's work. In analytical screening, callus is cut into two pieces. One half is subcultured, while from the other half, secondary metabolite is extracted and determined analytically by HPLC/GCMS/ELISA/Radioimmunoassay. Extraction is laborious and it limits number of the callus pieces that can be assayed to 200/d.

■■■■

16 ISOLATION OF MUTANTS FROM CELL CULTURE

16.1. INTRODUCTION

Successful regeneration of the plants from tissue, cell and protoplast cultures has led to the induction, screening and isolation of the mutants using the *in vitro* techniques.

16.1.1. Type of Culture and its Suitability for *In Vitro* Mutant Production

A variety of cultures – tissue, cell and protoplasts – may be used for *in vitro* isolation of the mutants.

Limitations (disadvantages) of callus culture for *in vitro* mutagenesis

Callus growing on solid culture medium is not considered as suitable for mutant induction or isolation because of the following reasons:

- The cells of the callus grow more slowly than suspension-cultured cells.
- The cells are not uniformly exposed to any mutagen or selective agent i.e., cells at the surface will be more exposed.
- As the cells are growing in large aggregates, mutated cells or resistant/tolerant cells that are deeply situated or surrounded by normal cells may not get a chance to regenerate.

Limitations of protoplasts for *in vitro* mutagenesis

- Techniques of protoplast isolation and culture are not available in wide number of plants.

- Isolation of large quantities of the protoplasts may prove very costly.
- Protoplasts are very fragile due to their extreme sensitivity to osmotic imbalances, and hence, are relatively difficult to handle.

Due to these limitations, protoplasts have been rarely used, that too in a few systems such as tobacco and *Petunia* only for *in vitro* selection of the mutants. However, protoplasts offer an advantage that they form reliable single cell culture systems, and hence, the possibility of selected calli/regenerants not being chimeric in nature is more.

Limitations and advantages of cell suspension cultures

Cell suspension cultures show faster growth rate than callus and can be manipulated similar to the culture of microorganisms. The most common limitation is frequent presence of the cell aggregates unlike the protoplasts. However, this limitation can be overcome by filtering the cultures through a carefully-chosen mesh (200–400 µm) to increase the proportion of single cells or a few-celled aggregations.

16.1.2. General Methodology for In Vitro Induction and Selection of Mutants using Suspension Cell Cultures

1. Select the cell culture, which has high regenerating ability.
2. Subculture the cells such that a high density of the single cells are available in exponential growth phase.

   This is important because when large number of the cells exists in exponential phase, they are actively dividing, and the possibility of mutagenesis during active DNA synthesis period ('S' phase of interphase) is more.
3. Concentrate the cells through centrifugation and expose them to physical mutagenic agents such as X-rays, UV etc. If chemical mutagens such as EMS or DES etc. are used, make suitable concentrations of the chemical in culture media and incubate the cells on shaker for the required period.
4. Wash the cells with fresh medium; this step may not be necessary if physical mutagens are used.
5. Grow the cells in dark for few replication cycles either before or after plating them into a suitable solid medium.

6. Grow the cells for several generations to allow segregation and expression of recessive mutations before actual selection is undertaken.

Method of selection

Most of the mutants have been selected by plating the suspension of the cells following mutagenesis on a selective medium with (or without) a particular chemical; the tolerance/ resistance of which is under study. Only those cells that can tolerate the presence/ absence of that chemical will survive and develop into calli while others will not survive.

16.1.3. Types of Mutants Selected using *In Vitro* Cultures

Amino acid and amino acid analogue resistance

The majority of the mutants isolated from *in vitro* cultures of higher plants have been selected for resistance to amino acid analogues or growth inhibitors. Most of the mutants have been isolated as resistant to either aromatic or lysine and threonine or their analogues (s-amino ethylcysteine – hydroxyl lysine). Resistance to antimetabolites can arise through one or more of the following several mechanisms:

- Reduced uptake of the analogue by changing the membrane permeability
- Decreased sensitivity of an amino acid biosynthetic enzyme to feedback inhibition leading to overproduction e.g., 5-methyl tryptophan resistance in *N. tabacum* through resistance of the first enzyme in the tryptophan pathway, (anthranilate synthetase) to feedback inhibition by tryptophan or its analogue (Widholm, 1972).
- Overproduction of naturally-occurring amino acid
- Excess of amino acid diluted effect of analogue in protein synthesis
- Detoxification of the compound through degradation e.g., paraflurophenylalanine can be degraded by phenylalanine ammonia lyase (Berlin 1980).
- Conversion of the excess chemical into other forms (secondary metabolites such as phenolic compounds)

For example, S-aminoethylcysteine resistance in *Arabidopsis thaliana* induced by EMS, hydroxyproline resistance in *Hordeum vulgare* induced by sodium azide, valine resistance induced in haploid *Nicotiana tabacum* by UV.

Antibiotic resistance

This is most extensively used in the analysis of somatic hybrids. They act mostly in the translational machinery such as ribosomes; 70s ribosomes are mostly sensitive. Since these types of ribosomes are present in the chloroplasts and mitochondria of higher plants, antibiotic resistance is expected to be cytoplasmically-inherited. However, some cases are known where they are inherited in a Mendelian way.

For example, streptomycin resistance and kanamycin resistance in tobacco (Maliga, 1973) showed maternal inheritance – a ribosomal protein was found to be different, chloramphenicol resistance in haploid tobacco, cycloheximide resistance in *Daucus carota* and *N. tabacum*, colchicine resistance in *Acer pseudoplatanus*.

Mutants affecting nitrogen metabolism

Nitrogen metabolism in higher plants involves several steps. Most of the mutants affecting nitrogen metabolism affect either nitrate uptake mediated by permease or reduction of nitrate to nitrite mediated by NR. In the absence of NR activity, chlorate is not converted to toxic chlorite; hence NR-deficient cell lines are chlorate-resistant.

Therefore, chlorate resistance has been used to identify NR-deficient mutants. The NR-deficient mutants are of two types: 'cxy' lines have cyt. c reductase activity, but not NR while 'nia' lines have xanthine dehydrogenase activity as well as lack cyt. c and NR activity. Complementation between 'cxy' and 'nia' lines through crossing or somatic hybridization restores the NR activity. This indicates that 'cnx' and 'nia' lines affect two parts of the NR enzyme. Chlorate-resistant mutants have also been reported in *Datura innoxia*, *Rosa damascena* and *Hyoscyamus muticus*.

Agriculturally-useful mutants

Various mutants in the cell cultures of higher plants have been selected that may have direct application in agriculture.

■■■■

17 PRACTICALS

EXPERIMENT NO. 1: MEDIA PREPARATION

Principle

In vivo plant cells, tissues and organs get their appropriate nutrients and growth requirements from the intact plant body for their organized growth and development. Isolated cells, tissues and organs also need nutrients for their *in vitro* growth and development. Hence, nutrients are supplied artificially according to the medium formulated by several workers.

Preparation of stock solutions

Preparation of MS micro stock I (40 X)

Components	mg/100 ml
H_3BO_3	248
KI	33.2
$Na_2MoO_4.\ 2H_2O$	10

Preparation of MS micro stock II (40 X)

Components	mg/100 ml
$MnSO_4.\ 4\ H_2O$	892
$ZnSO_4.\ 7\ H_2O$	344
$CuSO_4.\ 5\ H_2O$	1

Preparation of MS micro stock III (40 X)

Components	mg/100 ml
$FeSO_4 . 7 H_2O$	1112
Na_2EDTA	1492

Dissolve $FeSO_4 . 7 H_2O$ and Na_2EDTA separately in few millilitres of the boiling distilled water and mix. Heat it until the colour of solution changes to golden yellow. Store it in amber-coloured bottle.

Preparation of MS micro stock IV (40 X)

Components	mg/100 ml
Thiamine HCl	20
Pyridoxine HCl	20
Nicotinic acid	20
Glycine	80

Preparation of PGR stock solution (10 mg/100 ml)

Dissolve 10 mg of auxins/cytokinins in 100 ml of double-distilled water using appropriate solvent.

Protocol

Preparation of 1 l MS medium

1. Take 250 ml of double-distilled water and dissolve the following macrosalts one by one:

Macrosalts	mg/l
$MgSO_4 . 7 H_2O$	370
KH_2PO_4	170
KNO_3	1900
$NH_4 NO_3$	1650
$CaCl_2 . 2H_2O$	440
Inositol	100

2. Add 20 g of sucrose.
3. Add:

MS microstock I	2.5 ml
MS microstock II	2.5 ml
MS microstock III	2.5 ml
MS microstock IV	2.5 ml

4. Make up to 500 ml with double-distilled water.
5. Dispense 50 ml of above solution into 10 different beakers.
6. Add appropriate amount of PGR from the stock to get desired concentration.
7. Adjust pH to 5.6–5.8
8. Make up to 100 ml in each beaker with double-distilled water.
9. Add 800 mg of agar to each beaker.
10. Melt the agar in micro oven with intermittent shaking.
11. Dispense the medium into the culture tubes.
12. Close the culture tubes with either cotton plugs or aluminium foil.
13. Autoclave for 15 min under 15 lbs/in^2 at 121°C temperature.
14. Store in the culture room until use.

EXPERIMENT NO. 2: RAISING OF ASEPTIC SEEDING

Principle

Surface sterilization of the plant tissue may cause some deleterious effects because most of the sterilants are toxic chemicals. The seeds can more or less resist such deleterious effects due to the presence of its seed coat. Hence, to avoid the surface sterilization of the plant tissue, seeds are surface-sterilized and are cultured on simple basal nutrient medium. The seeds in the culture germinate and give rise to an aseptic seedling. Explants from such seedling, grown under aseptic and controlled conditions, are the most suitable material for culture and need no further surface sterilization.

Protocol

1. Wash the dry seeds with tap water.
2. Treat the seeds with 4–5 drops of Tween-20 for 30 min.
3. Wash the seeds with running tap water.
4. Carry the remaining procedure in laminar flow cabinet.
5. First, wash the seeds with sterile distilled water followed by a treatment with surface sterilant such as mercuric chloride, hydrogen peroxide, sodium hypochlorite etc. for a few minutes.
6. Decant the surface sterilant and rinse 4–5 times with sterile distilled water.
7. Rinse quickly with 70% ethanol (i.e., 60 s).
8. Wash with sterile distilled water 4–5 times.
9. Now, the seeds are ready for inoculation.
10. Open the closure of the culture vial containing the BM.
11. Flame the neck of the culture vial and transfer the seed in quick succession on the the medium.
12. Replace the closure.
13. Incubate the seed at 25 ± 2°C with a photoperiod of 16 h light and 8 h dark.
14. Take the observations every week.

EXPERIMENT NO. 3: CALLUS CULTURE

Principle

Three important criteria should be accomplished for a successful initiation of the callus culture:

1. Aseptic preparation of the plant material
2. Selection of suitable nutrient medium supplemented with appropriate ratio of auxins and cytokinin or only appropriate auxin
3. Incubation under controlled condition

Protocol

1. Aseptically-germinated seedling will be free of contaminants. This can be directly used as an explant.
2. Transfer the plant from the culture tube to a sterile petri plate containing filter paper.
3. Excise the root, internode, hypocotyls, leaf and cotyledon, and transfer them into another sterile petri plate.
4. Transfer each explant separately into appropriate concentration using sterile foreceps, and close the culture tube with aluminium foil.
5. Incubate the culture tubes at 25 ±2°C with a photoperiod of 16 h light and 8 h dark.
6. Take the observations every week.

EXPERIMENT NO. 4: SUBCULTURING OF CALLUS

Principle

Callus is homogenous, unorganized mass of cells produced due to the multiplication of the explant on the medium. This callus can be subcultured for further studies such as organogenesis, somatic embryogenesis, cell suspension culture etc. For this purpose, callus is induced to produce more calli on a medium supplemented with growth hormones.

Protocol

1. Prior to subculturing, undertake all the necessary steps to maintain aseptic conditions.
2. Since callus is produced under aseptic conditions, surface sterilization is not necessary, but maintenance of the aseptic conditions during inoculation is necessary.
3. If the callus is of sufficient size in the culture tube, directly subculture it on to a fresh medium.
4. Hold the callus-containing tube and the fresh-medium containing tube near the flame, and hold the sterile foreceps in the right hand.
5. Keep the tubes in an inclined position, open and sterilize the open-end by passing through the flame.

6. Immediately after deplugging, insert the sterilized cool foreceps into the callus-containing culture tube. Take out the callus using the foreceps and immediately transfer the callus to a fresh medium tube. Close the tube with aluminium foil.
7. If the callus is too large, in order to subculture, first transfer the callus to a sterile petri plate containing filter paper. Then cut the callus into small bits with a sterile blade.
8. Transfer the small bits into fresh medium, close the tubes and label properly.
9. Carry out subculturing every 4 wk to get more amount of callus for further studies.

EXPERIMENT NO. 5: MULTIPLE SHOOT INDUCTION

Protocol

1. Aseptically-germinated seedling will be free of contaminants. This can be directly used as an explant.
2. Transfer the plant from culture tube to a sterile petri plate containing filter paper.
3. Excise the nodes, cotyledonary node, shoot-tip, and transfer them into another sterile petri plate.
4. Transfer each explant separately into appropriate concentration using sterile forecep and close the culture tube with a cotton plug or aluminium foil.
5. Incubate the culture tubes at 25 ± 2°C with a photoperiod of 16 h light and 8 h dark.
6. Take the observations every week.
7. Maintain the polarity of the explants during the inoculation.

EXPERIMENT NO. 6: SHOOT-TIP CULTURE FROM MATURE EXPLANT

Principle

The excised shoot-tips can be cultured aseptically on an agar-solidified simple nutrient medium or liquid medium under appropriate conditions, and they will grow out small leafy shoots or multiple shoots.

Alternatively, small callus at its cut-ends results in development of large number of shoot primordia. These shoot primordia grow out into multiple shoots. Once the shoots have been grown directly from an excised shoot-tip, they can be further propagated.

Protocol

1. Collect actively growing shoot-tips and wash with tap water.
2. Treat the explants with 4–5 drops of Tween-20 for 30 min.
3. Wash them with running tap water until the traces of Tween-20 are removed.
4. Carry out the remaining procedure in laminar flow cabinet.
5. First, wash the explants with sterile distilled water, followed by a treatment with surface sterilant such as mercuric chloride, hydrogen peroxide, sodium hypochlorite etc. for a few minutes.
6. Decant the surface sterilant and rinse 4–5 times with sterile distilled water.
7. Rinse quickly with 70% ethanol (i.e., for 60 s).
8. Wash with sterile distilled water 4–5 times
9. Now, transfer the explant on to a sterile petri plate containing filter paper.
10. Trim the shoot-tip and transfer it to another petri plate.
11. Flame the closure of the culture vial and transfer the explant on to the medium with the help of sterile foreceps. Flame the foreceps before and after use.
12. Close the culture tube with a cotton plug or aluminium foil.
13. Incubate the culture tubes at 25 ± 2°C with a photoperiod of 16 h light and 8 h dark.
14. Take the observations every week.

EXPERIMENT NO. 7: SUBCULTURING FOR MULTIPLICATION AND ROOTING

Protocol

1. As the shoots are grown aseptically on the culture medium, there is no need of surface sterilization.
2. Transfer the multiple shoot culture on to a sterile petri plate containing filter paper.
3. Excise the shoots with 4–5 nodes using scalpel and place them on a rooting medium with sterile foreceps.
4. Subculture the remaining culture for the growth of other shoots.
5. Place the shoots kept for rooting, in darkness for 48 h and then transfer them into light with photoperiod of 16 h light and 8 h dark.

EXPERIMENT NO. 8: ACCLIMATIZATION

Principle

Micropropagation, on a large scale, can be successful only when the plants show high survival rate after transfer from culture tubes to the soil, and the cost involved in the process is low.

1. Take out *in vitro* raised plantlets with well-developed roots from the test tubes/bottles.
2. Clean the plantlets with sterilized double-distilled water to remove all the traces of agar.
3. Plant these plantlets in earthenware pots containing sterilized vermiculite: sand (3:1).
4. Cover the plantlets with big polythene bag and incubate at 25 ± 2ºC for 20 d.
5. Add liquid quarter-strength MS BM devoid of sucrose instead of water.
6. Make small perforations on the polythene bag to reduce the relative humidity.

7. Slowly, the width of the holes is increased until the relative humidity inside the polythene bag and outside the chamber becomes equal.
8. After this, remove polythene bag and directly expose the pots to the controlled temperature (25 ± 2°C).
9. Gradually, transfer the pots to a room having diffused light.
10. Eventually, shift the plants to a room having sunlight.

EXPERIMENT NO. 9: PRODUCTION OF ARTIFICIAL SEEDS

Principle

The concept of artificial seeds has been developed from somatic embryos, which are formed adventitiously from *in vitro* cultured somatic tissue. Somatic embryos are not enclosed by the seed coats and do not survive if these are sown directly into the field soil. Hence, tissue culturists tried to develop a technique by which isolated somatic embryos could be encapsulated by a protective gel-like substance. As a result, embryoids are able to survive and would not desiccate even after transfer to soil. Such encapsulated somatic embryos could be used as artificial seeds. Alternately, synthetic seeds can also be produced using other propagules such as shoot-tip and nodes of *in vitro* multiple shoots.

Protocol

1. Prepare MS medium by adding macronutrients and micronutrients.
2. Add 3% sodium alginate (hydrogel) instead of agar agar.
3. Prepare calcium chloride solution.
4. Dispense sodium alginate + MS medium and calcium chloride in separate flasks.
5. Blot the axillary buds with nodal portion (0.5 cm length) or shoot-tips from regenerated shoots or somatic embryos from cultured callus to dryness on a sterile filter paper.
6. Then mix these explants in 3% sodium alginate + MS medium supplemented with 2% sugar.